10 Nebraska NSCAS Grade 6 Math Practice Tests

The Ultimate Test Prep Collection with Answer Explanations

Dr. A. Nazari

10 Practice Tests

The Grand Championship Collection

Welcome, future Math Champion!

You hold the **ultimate collection** —
ten full-length practice tests designed to take you
from first attempt to **complete mastery.**

- Conquer every Grade 6 topic
- Build unshakeable confidence
- Rise from Bronze to Gold to Champion
- Arrive at test day fully prepared

The championship begins now.

★ ★ ★ ★ ★

> **"** Ten tests may seem like a marathon, but champions are made one step at a time. Trust the process! **"**

 The Champion's Path

Your 4-phase journey from Bronze to Champion

Bronze Round (Tests 1–3)

Your warm-up matches. Take these **untimed** to learn the format and set your baseline. Read the answer explanations after each test — this is where you build your foundation.

Silver Round (Tests 4–6)

Set a timer for **75 minutes**. Focus on the topics that tripped you up in Bronze. Practice showing your work on every problem. Your accuracy should be climbing.

Gold Round (Tests 7–9)

Full timed conditions (**60 minutes**). Simulate the real exam environment. Review only the questions you missed — targeted practice is the key to gold.

Championship Final (Test 10)

Your final match. Full exam conditions — timed, quiet, no breaks. This is your victory lap. Show yourself how far you've come!

Your Championship Kit

- **10 Full-Length Practice Tests** — every Grade 6 topic
- **Formula Reference Sheet**
- **Complete Answer Key** with explanations
- **Championship Scoreboard** to track your rise

 **Champion's Tip:** Space your tests 2–3 days apart. Use the days in between for targeted review. By Test 10, you'll be amazed at your transformation.

👑 The Champion's Playbook 👑

Seven rules that separate champions from the rest

I **Read every question twice.** The first read tells you the topic. The second tells you exactly what to solve for. Champions never skim.

II **Mark the clues.** Circle key numbers, underline the question, and cross out information that's just there to distract you.

III **Choose your strategy.** Before touching pencil to paper, decide: Am I setting up a ratio? Solving an equation? Finding area? Name the approach.

IV **Solve, then match.** For multiple choice — work the problem on scratch paper first, then find your answer among the choices.

V **Eliminate and conquer.** Cross out obviously wrong answers. If you're left with two, you've already doubled your odds. Make an educated pick.

VI **Estimate to verify.** After solving, ask: "Is this answer reasonable?" A quick mental estimate catches most calculation errors.

VII **Leave nothing blank.** Even a well-reasoned guess is worth more than an empty space. Use partial work to support your answer.

🕐 Timing Mastery

Tests 1–3: **Untimed** (build foundation) ＞ Tests 4–6: **75 min** (build speed) ＞ Tests 7–10: **60 min** (championship conditions)

⭐ Grade 6 Championship Topics

🏅 Ratios & Proportions 🏅 Integers & Rational Numbers 🏅 Expressions & Equations

🏅 Geometry & Measurement 🏅 Statistics & Data Analysis

*A true champion isn't someone who never makes mistakes — it's someone who learns from **every single one**. After each test, review your errors carefully. That's where the real growth happens.*

Find more at
ViewMath.com/NE-Grade6

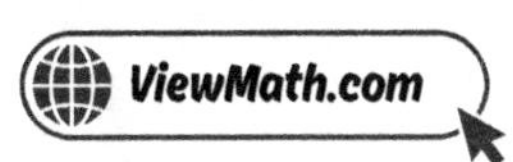

👑 *The Champion's Toolkit* 👑

🏆 Required Equipment

✏️	**Sharpened Pencils**	Two #2 pencils — champions always have a backup
◢	**Quality Eraser**	A clean, soft eraser that won't smudge your work
▣	**Scratch Paper**	Blank paper for calculations, diagrams, and number lines
📏	**Ruler**	Essential for geometry and coordinate plane questions
⏱	**Timer**	Begin using from the Silver Round onward
🔊	**Quiet Workspace**	A calm, well-lit area free from distractions

⚕ Permitted in Competition

- ✔ Pencil and eraser
- ✔ Scratch paper (provided)
- ✔ Ruler (if specified)
- ✔ Formula reference in this book

🚫 Not Permitted

- ✖ Calculators
- ✖ Electronic devices
- ✖ Textbooks or notes
- ✖ Outside help

👥 For Parents & Teachers

- With 10 tests, space them **2–3 days apart**. This gives time to review mistakes and study between rounds.

- Let your child take Tests 1–3 untimed to build familiarity and establish a baseline.

- After each test, go through the Answer Key together. Focus on **understanding the reasoning**, not memorizing answers.

- Use the Championship Scoreboard to visualize long-term progress. Celebrate improvements at every tier!

- Pair with our **Grade 6 Math Study Guide** for topics that need sustained attention.

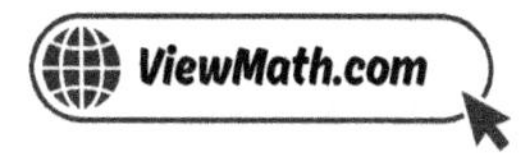

Formula Reference Sheet

◢◣ Area Formulas

Rectangle	$A = l \times w$
Parallelogram	$A = b \times h$
Triangle	$A = \dfrac{1}{2} \times b \times h$
Trapezoid	$A = \dfrac{1}{2}(b_1 + b_2) \times h$

⬢ Volume

Rectangular Prism $V = l \times w \times h$

⬢ Surface Area

Find the area of each face, then add them all up.

Rectangular Prism:

$SA = 2lw + 2lh + 2wh$

⬇ Order of Operations

P Parentheses first

E Exponents

M/D Multiply & Divide (left to right)

A/S Add & Subtract (left to right)

％ Ratios & Percents

Ratio: $a : b$ or $\dfrac{a}{b}$

Unit rate: amount per 1 unit

Percent: a ratio out of 100

Part = Percent × Whole

⚖ Integers & Absolute Value

Integers:

$\ldots, -3, -2, -1, 0, 1, 2, 3, \ldots$

$|-5| = 5 \quad |5| = 5$

Absolute value = distance from 0

X¹ Expressions & Equations

Exponent: $3^4 = 3 \times 3 \times 3 \times 3 = 81$

Variable: a letter that stands for a number

Equation: two expressions joined by =

Inequality: uses $<, >, \leq, \geq$

⊕ Coordinate Plane

Ordered pair: (x, y)

x-axis: horizontal y-axis: vertical

Origin: $(0, 0)$

Four quadrants (I, II, III, IV)

📊 Statistics

Mean: sum of values ÷ count

Median: middle value (sorted)

Range: max − min

Championship Scoreboard

Track your rise through every championship round

Champion's Name: _______________________________

Round	Tier	Date	Score	Rating
1	Bronze		/	
2	Bronze		/	
3	Bronze		/	
4	Silver		/	
5	Silver		/	
6	Silver		/	
7	Gold		/	
8	Gold		/	
9	Gold		/	
10	♔		/	

× ____

My strongest topics (where I consistently score well):

____

____

Topics I improved on the most from Bronze to Gold:

____

____

My score trend (Bronze avg → Gold avg → Championship):

____

____

One strategy that helped me improve the most:

____

____

My confidence level for the real test (1–10): _________ / 10

Find more at
ViewMath.com/NE-Grade6

Continue Learning at ViewMath Academy!

For Parents, Teachers & Students

Great job on the practice tests! Want to keep improving? ViewMath Academy is your **free online companion** to this book.

- 📊 **Score Analyzer** — Enter your answers and instantly see which topics need more practice

- 📖 **Interactive Lessons** — Review the concepts behind each question with clear explanations

- 🧩 **Adaptive Quizzes** — Practice your weak topics with questions that match your level

- 📈 **Progress Tracking** — See your mastery grow across all Grade 6 math topics

- 🎓 **Personalized Dashboard** — A learning plan tailored just for you

Scan to visit ViewMath Academy

ViewMath.com/NE-Grade6

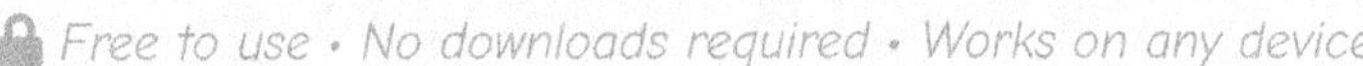 🔒 Free to use · No downloads required · Works on any device 

⭐ Table of Contents ⭐

Here's what we'll explore together!

 Let's learn and have fun!

Practice Test 1

 30 Questions

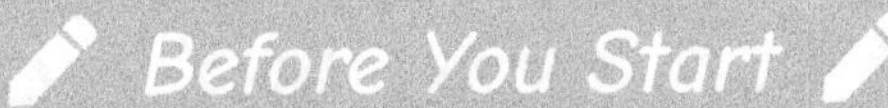

✏ Before You Start ✏

- ✔ **Read each question carefully** before choosing your answer.
- ✔ **Show your work** on scratch paper when you need to.
- ✔ **Skip hard questions** and come back to them later.
- ✔ **Check your answers** when you're done.
- ✔ **Take your time** — there's no rush!

⭐ You've Got This! ⭐

Do your best and show what you know!

1. A garden has flowers and vegetables in a ratio of 4 : 1. There are 20 flowers. How many vegetables are there?

Your Answer:

2. The graph below shows the distance traveled by a delivery truck over time.

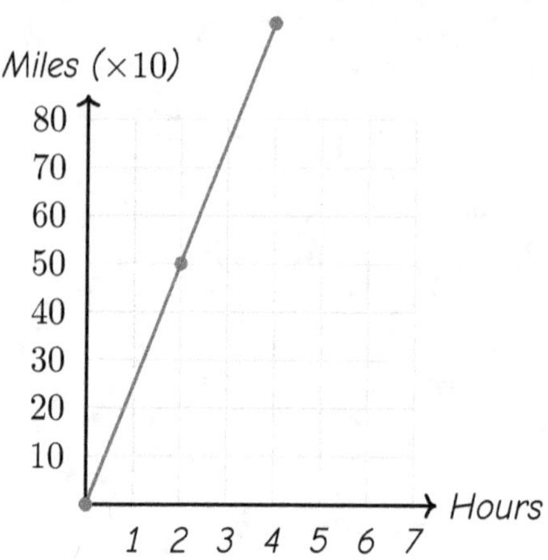

What is the truck's unit rate in miles per hour?

(A) 50 miles per hour

(B) 20 miles per hour

(C) 25 miles per hour

(D) 10 miles per hour

3. The double number line below shows equivalent ratios of scoops of mix to cups of water.

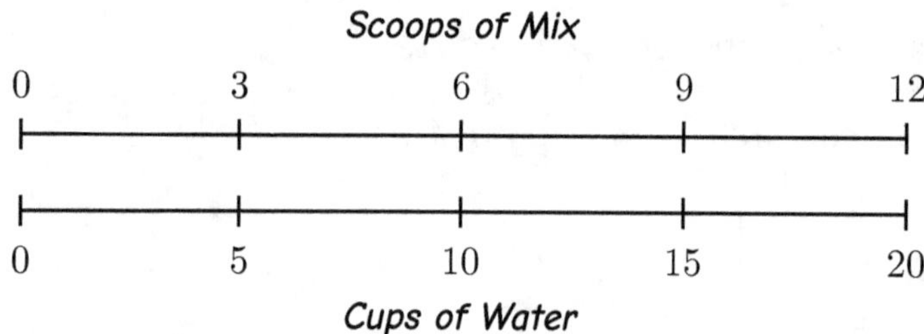

How many cups of water go with 9 scoops of mix?

(A) 10

(B) 12

(C) 15

(D) 18

Find more at
ViewMath.com/NE-Grade6

ViewMath.com

4. A graph of a ratio relationship passes through $(0, 0)$ and $(5, 20)$. What is the unit rate?

(A) 5

(B) 20

(C) 4

(D) 25

5. The double number line below shows a map scale.

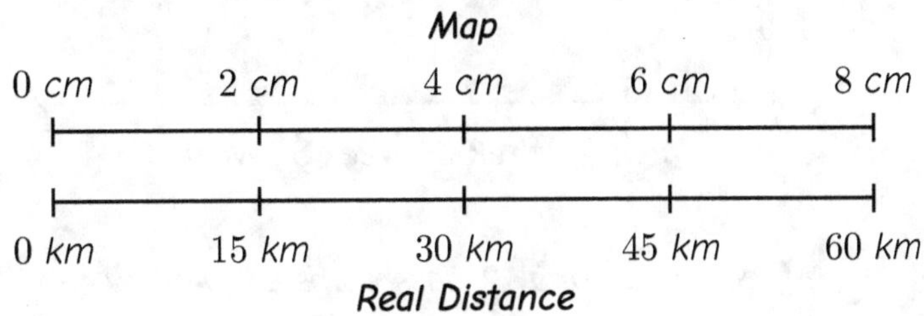

Two cities are 5 cm apart on the map. What is the real distance?

(A) 25 km

(B) 30 km

(C) 37.5 km

(D) 35 km

6. A bottle holds 750 mL. How many full bottles can you fill from a 6-liter jug?

Your Answer:

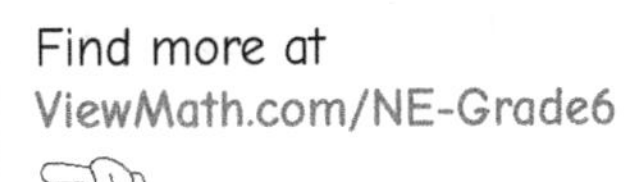

7. The bar graph below shows the number of cans a school collected for recycling each week.

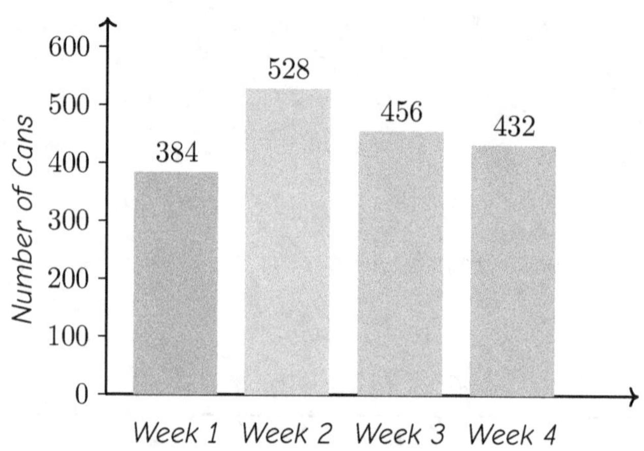

The school wants to divide all the cans equally among 12 classrooms. How many cans does each classroom receive?

(A) 180

(B) 1,800

(C) 150

(D) 130

8. A water jug holds 5.4 liters. How many liters are in 6 jugs?

(A) 32.4

(B) 324

(C) 3.24

(D) 30.4

9. Find $|-9| + |-6|$.

Your Answer:

10. The point $(-7, 3)$ is reflected across the x-axis. Write the coordinates of the reflected point.

Your Answer:

11. A store charges the prices shown below. Which expression gives the total cost for s sandwiches and d drinks?

Sandwich $6 each	Drink $2 each

A $6 + 2$ B $6s + 2d$

C $8sd$ D $2s + 6d$

12. A student made a factor tree for the expression $4(x + 3)$, shown below. Fill in the missing pieces labeled A and B.

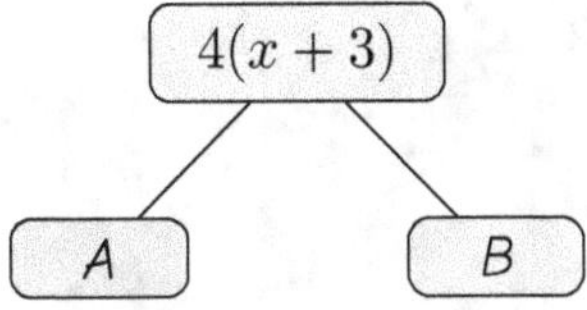

Your Answer:

13. Evaluate $3(p + 2) - p$ when $p = 8$.

Your Answer:

14. Which expression is equivalent to $3(x + 5)$?

A $3x + 5$ B $3x + 15$

C $8x$ D $3x + 8$

15. Five friends share the cost of a gift equally. Each friend pays $9. What is the total cost of the gift?

Your Answer:

Find more at
ViewMath.com/NE-Grade6

ViewMath.com

16. *The temperature stayed below 0 degrees. Which inequality represents the temperature t?*

(A) $t > 0$

(B) $t < 0$

(C) $t \leq 0$

(D) $t \geq 0$

17. *A student says the graph of $x < 8$ and the graph of $x \leq 8$ look exactly the same. Is this correct?*

Your Answer:

18. *Complete the table for $y = 4x - 1$:*

$x = 1$: $y = ?$ $x = 3$: $y = ?$ $x = 5$: $y = ?$

Your Answer:

19. *A triangular sail has a base of 3 m and a height of 7 m. How much fabric is needed to make 2 identical sails?*

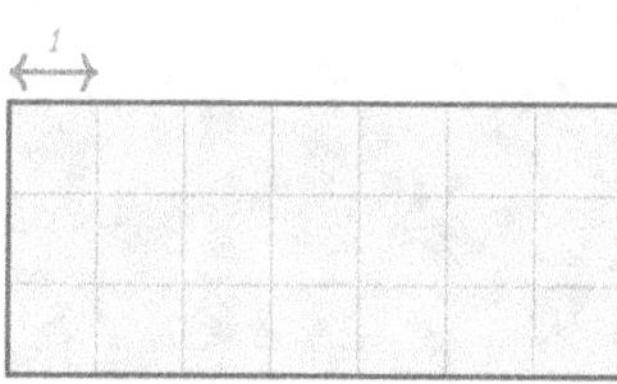

(A) $10.5\ m^2$

(B) $42\ m^2$

(C) $21\ m^2$

(D) $20\ m^2$

20. *Two boxes have the same volume. Box A is $6 \times 4 \times 5$. Box B is $10 \times 3 \times h$. What is the height h of Box B?*

(A) 2

(B) 4

(C) 6

(D) 8

Find more at
ViewMath.com/NE-Grade6

ViewMath.com

21. Points $A(0, 4)$ and $B(0, -5)$ are connected by a line segment. What is the length of $\overline{AB}$?

(A) 1 unit

(B) 5 units

(C) 4 units

(D) 9 units

22. A right triangle has vertices $(-4, 0)$, $(2, 0)$, and $(-4, 5)$. What is the area?

(A) 30 square units

(B) 15 square units

(C) 11 square units

(D) 20 square units

23. A rectangular prism has dimensions 10 m, 3 m, and 7 m. What is its surface area?

(A) 210 m^2

(B) 242 m^2

(C) 262 m^2

(D) 121 m^2

24. A circular pool has a diameter of 20 feet. What is the area of a pool cover that fits exactly over the pool? Use $\pi \approx 3.14$.

(A) 62.8 ft^2

(B) 157 ft^2

(C) 314 ft^2

(D) 1,256 ft^2

25. When data is "skewed to the right," what does that mean?

(A) Most values are on the left, with a tail stretching to the right.

(B) Most values are on the right, with a tail stretching to the left.

(C) The data is symmetric.

(D) All values are equal.

Find more at
ViewMath.com/NE-Grade6

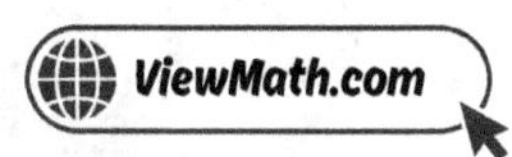

26. A frequency table shows how many books students read last month:

Books	Students
1	3
2	5
3	4
4	2
5	1

What is the median number of books?

(A) 2

(B) 2.5

(C) 3

(D) 4

27. Data (in order): $3, 5, 7, 8, 10, 12, 14$. What is the IQR?

(A) 5

(B) 7

(C) 9

(D) 11

28. A box plot shows: $min = 25$, $Q1 = 35$, $median = 50$, $Q3 = 65$, $max = 80$. Find the range and IQR.

Your Answer

29. Two classes both have a mean of 75. Class A: $MAD = 3$. Class B: $MAD = 9$. A student from each class is picked at random. Whose score is more likely to be close to 75?

(A) The student from Class A

(B) The student from Class B

(C) Both equally likely

(D) Cannot determine

Find more at
ViewMath.com/NE-Grade6

30. A spinner has 8 equal sections. Three sections are blue and five sections are red. What is the probability of **not** landing on blue? Write your answer as a decimal.

Your Answer:

Find more at
ViewMath.com/NE-Grade6

 End of Practice Test 1 

Great job finishing the test!

 My Score

I got _____________ out of 30 questions right.

Check your answers in the **Answer Key** at the back of the book.

Review any questions you missed. That's how we learn!

Check Your Score Online!

Visit **ViewMath Academy** to enter your answers and see which topics you need to review. You can also explore lessons, take quizzes, track your scores, and save your progress!

viewmath.com/score/6.1.NE.16

Or go to viewmath.com/score and enter code: 6.1.NE.16

2

Practice Test 2

 30 Questions

 Before You Start 

- ✔ **Read each question carefully** before choosing your answer.
- ✔ **Show your work** on scratch paper when you need to.
- ✔ **Skip hard questions** and come back to them later.
- ✔ **Check your answers** when you're done.
- ✔ **Take your time** — there's no rush!

★ You've Got This! ★

Do your best and show what you know!

1. Look at the tape diagram below.

Boys: □□□□
Girls: □□□□□□

There are 24 girls. How many boys are there?

(A) 4

(B) 12

(C) 16

(D) 20

2. The table shows the earnings of two babysitters.

	Hours Worked	Earnings
Lily	5	$60
Noah	4	$52

Part A: Find each babysitter's unit rate (dollars per hour).

Part B: Who earns more per hour? How much more?

Your Answer:

3. The graph below shows points that represent equivalent ratios of flour to sugar in a recipe.

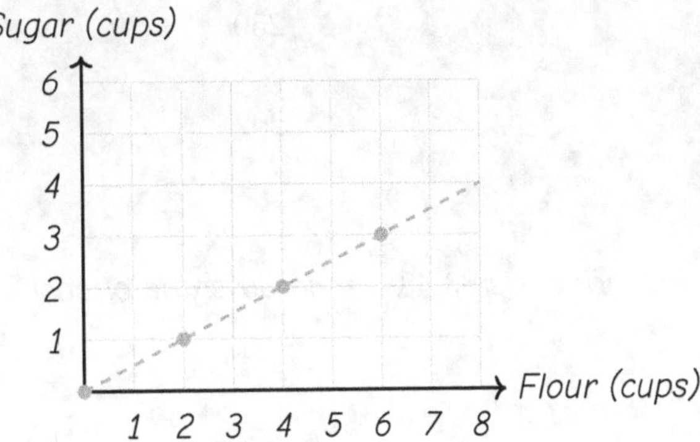

Part A: What is the ratio of flour to sugar?

Part B: If you need 8 cups of flour, how many cups of sugar do you need?

Your Answer:

---

4. The points $(2, 5)$ and $(4, 10)$ are on a graph. What is the ratio $x : y$?

(A) $5 : 2$ (B) $1 : 2$

(C) $2 : 5$ (D) $4 : 5$

---

5. A pool fills at a rate of 12 gallons per minute. How long does it take to fill 360 gallons?

(A) 25 minutes   (B) 30 minutes

(C) 36 minutes   (D) 40 minutes

---

6. A race is 10 kilometers. How many centimeters is that?

(A) 100,000   (B) 10,000

(C) 1,000,000   (D) 1,000

---

7. What is $756 \div 3$?

(A) 252

(B) 250

(C) 258

(D) 202

8. A book costs $12.95 and a bookmark costs $3.48. Jenna buys one of each. How much does she spend in all?

Your Answer:

9. What is $|-12|$?

(A) -12

(B) 12

(C) 0

(D) $-(-(-12))$

10. Explain why the point $(9, 0)$ is not in any quadrant.

Your Answer:

11. A baker makes c cupcakes. She packs them equally into 6 boxes. Which expression shows the number in each box?

(A) $c - 6$

(B) $6c$

(C) $c \div 6$

(D) $c + 6$

12. In the expression $5(y + 8)$, name the two factors.

Your Answer:

Find more at
ViewMath.com/NE-Grade6

ViewMath.com

13. The triangle below has a base of $b = 10$ cm and a height of $h = 6$ cm. Use the formula $A = \frac{1}{2} \times b \times h$ to find its area.

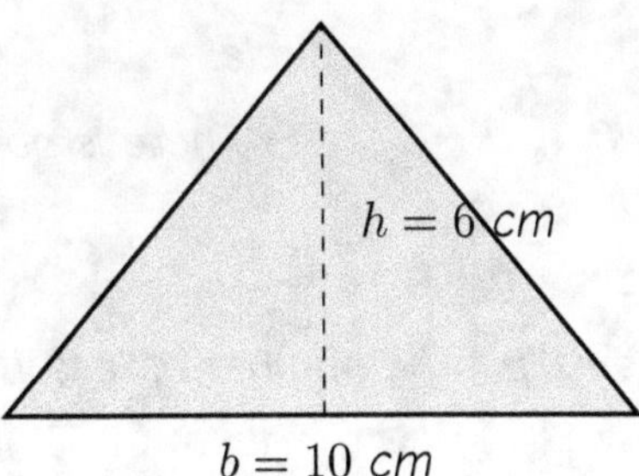

Your Answer:

14. Simplify: $3a + 7 + 5a - 2a + 3$

Your Answer:

15. Which equation has the solution $x = 12$?

(A) $x + 5 = 7$

(B) $3x = 36$

(C) $x - 4 = 16$

(D) $x + 12 = 12$

16. Which inequality represents "a number y is fewer than 20"?

(A) $y > 20$

(B) $y \geq 20$

(C) $y < 20$

(D) $y \leq 20$

17. What is the difference between the graphs of $x > 5$ and $x \geq 5$?

(A) They shade in different directions

(B) $x > 5$ uses an open circle; $x \geq 5$ uses a closed circle

(C) $x > 5$ shades right; $x \geq 5$ shades left

(D) There is no difference

18. Which equation represents this relationship: "The number of legs L is 4 times the number of dogs d"?

(A) $d = 4L$

(B) $L = d + 4$

(C) $L = 4d$

(D) $L = d \div 4$

19. A right triangle has legs of 9 in and 12 in. What is its area?

Your Answer:

20. A rectangular prism has a volume of 120 cm³. Its length is 10 cm and width is 4 cm. What is the height?

(A) 3 cm

(B) 12 cm

(C) 6 cm

(D) 30 cm

21. A rectangle on the coordinate plane has a length of 10 units and width of 4 units. One vertex is at the origin $(0,0)$ and the sides are along the axes. Which of the following could be the opposite vertex?

(A) $(10, 4)$

(B) $(14, 0)$

(C) $(4, 4)$

(D) $(10, 10)$

22. A square on the coordinate plane has vertices $(-3, -3)$, $(5, -3)$, $(5, 5)$, and $(-3, 5)$. What is the area?

Your Answer:

Find more at
ViewMath.com/NE-Grade6

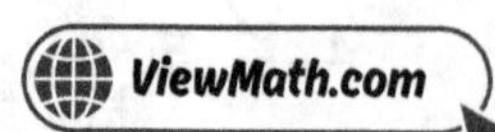

23. How many faces does a rectangular prism have?

(A) 4

(B) 5

(C) 6

(D) 8

24. A circular tablecloth has a diameter of 6 feet. Lace trim costs $2 per foot. How much will it cost to add lace trim around the entire edge of the tablecloth? Use $\pi \approx 3.14$

Your Answer:

25. Look at the dot plot below.

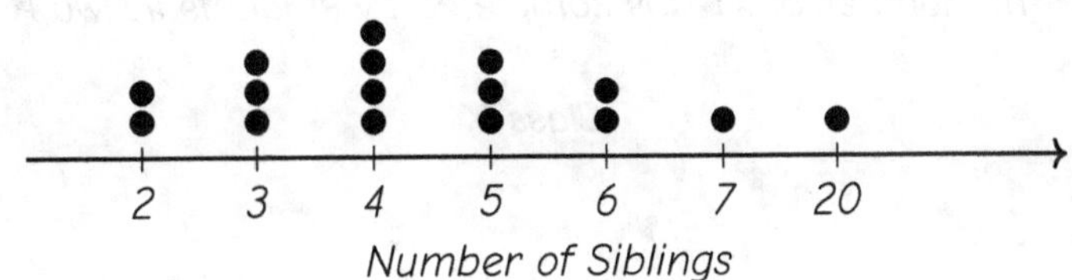

Which value is an outlier?

(A) 2

(B) 4

(C) 7

(D) 20

26. The dot plot below shows daily high temperatures (°F) for 7 days.

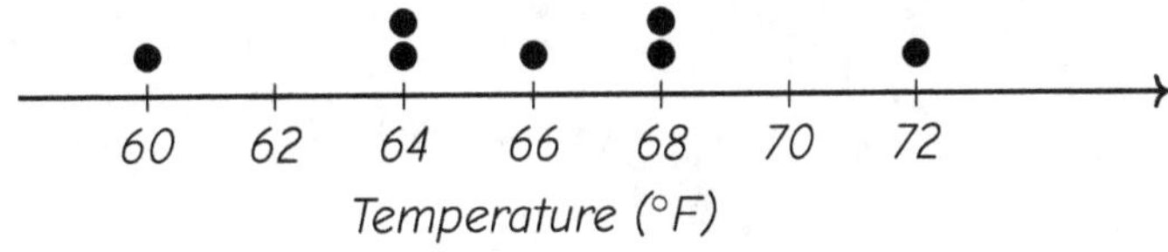

Find the mean and median temperature from the dot plot.

Your Answer:

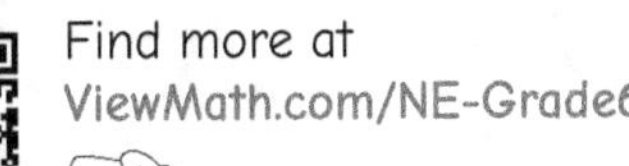

Find more at
ViewMath.com/NE-Grade6

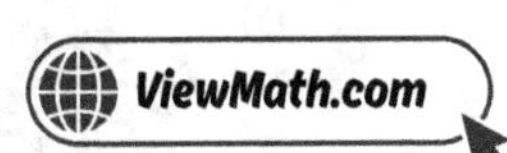

27. What is the range of the data set $8, 12, 15, 20, 25$?

(A) 8

(B) 13

(C) 17

(D) 25

28. Which data set has the five-number summary: $5, 10, 15, 20, 25$?

(A) $5, 10, 15, 20, 25$

(B) $5, 8, 10, 15, 18, 20, 22, 25$

(C) $5, 7, 10, 12, 15, 18, 20, 23, 25$

(D) Any of these could have that five-number summary.

29. The dot plots below show the number of sit-ups completed by students in two PE classes.

Class X

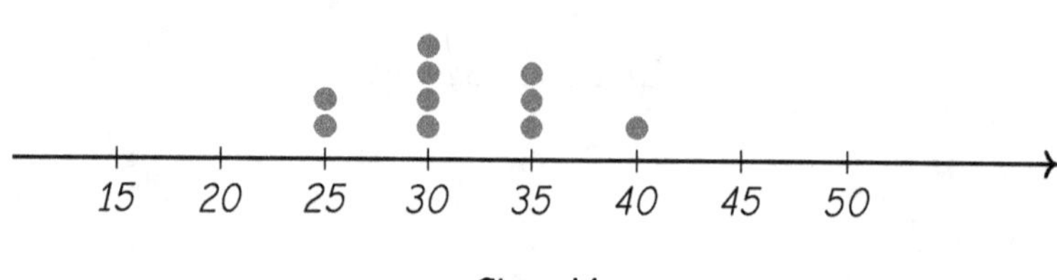

Class Y

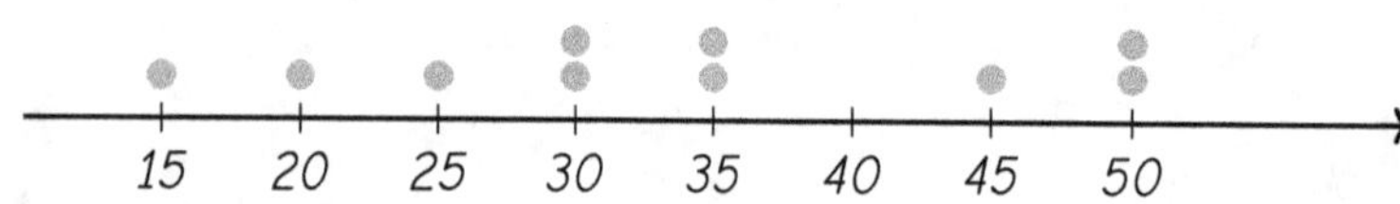

Which class has data that is more spread out?

(A) Class X

(B) Class Y

(C) They have the same spread.

(D) Cannot be determined.

Find more at
ViewMath.com/NE-Grade6

30. *The probability scale below shows four events. Which event has a probability closest to 0.75?*

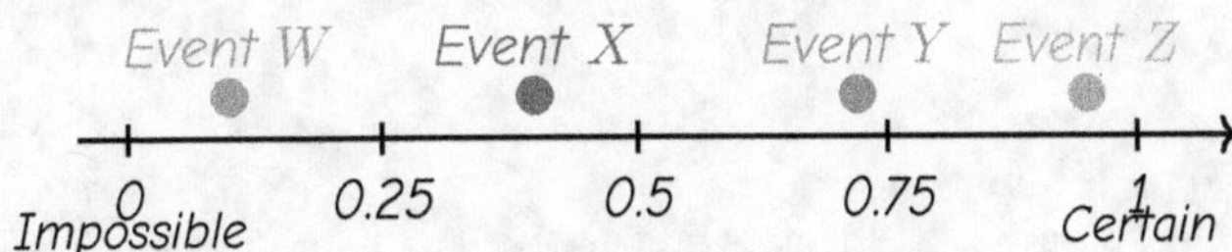

(A) Event W

(B) Event X

(C) Event Y

(D) Event Z

Find more at
ViewMath.com/NE-Grade6

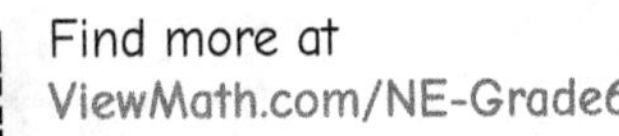

 # End of Practice Test 2 

Great job finishing the test!

☑ My Score

I got __________ out of 30 questions right.

*Check your answers in the **Answer Key** at the back of the book.*

💡 *Review any questions you missed. That's how we learn!*

📊 Check Your Score Online!

Visit **ViewMath Academy** to enter your answers and see which topics you need to review. You can also explore lessons, take quizzes, track your scores, and save your progress!

viewmath.com/score/6.1.NE.17

Or go to *viewmath.com/score* and enter code: *6.1.NE.17*

Practice Test 3

 *30 Questions*

✏️ Before You Start ✏️

- ✔ **Read each question carefully** before choosing your answer.
- ✔ **Show your work** on scratch paper when you need to.
- ✔ **Skip hard questions** and come back to them later.
- ✔ **Check your answers** when you're done.
- ✔ **Take your time** — there's no rush!

⭐ You've Got This! ⭐

Do your best and show what you know!

1. A baker uses 2 eggs **per** 3 cups of flour. Which ratio represents flour to eggs?

(A) $2:3$

(B) $3:5$

(C) $3:2$

(D) $2:5$

2. A box of 12 muffins costs $9. What is the cost per muffin?

(A) $1.25

(B) $0.75

(C) $1.33

(D) $3.00

3. A store sells 7 bananas for $2. How much do 21 bananas cost?

(A) $4

(B) $6

(C) $9

(D) $14

4. A ratio graph passes through $(6, 2)$. What is the value of y when $x = 15$?

(A) 3

(B) 5

(C) 7

(D) 10

5. A store sells 3 T-shirts for $24. How much do 10 T-shirts cost?

(A) $72

(B) $80

(C) $70

(D) $90

6. Convert 3 miles to feet. (1 mile $= 5{,}280$ feet)

(A) 5,280

(B) 10,560

(C) 15,840

(D) 21,120

Find more at
ViewMath.com/NE-Grade6

7. What is $7,344 \div 24$?

   (A) 36   (B) 360

   (C) 306   (D) 3,006

8. What is 1.2×0.3?

   (A) 3.6   (B) 0.36

   (C) 36   (D) 0.036

9. Which number has an absolute value of 6?

   (A) 6 only   (B) −6 only

   (C) Both 6 and −6   (D) No number has an absolute value of 6

10. Which sign convention describes all points in Quadrant II?

   (A) $(+,+)$ (B) $(-,+)$

   (C) $(-,-)$ (D) $(+,-)$

11. Sam has x stickers. He gives away 6. Which expression shows how many stickers Sam has now?

   (A) $x + 6$ (B) $6x$

   (C) $6 - x$ (D) $x - 6$

12. How many terms does the expression $4(a + b)$ have when written in its factored form?

Your Answer:

13. Evaluate $k^2 + k$ when $k = 3$.

   (A) 6   (B) 9

   (C) 12   (D) 15

14. Use the distributive property to expand $8(y + 3)$.

> *Your Answer:*

15. Solve: $12p = 84$

   (A) $p = 72$ (B) $p = 96$

   (C) $p = 7$ (D) $p = 6$

16. A library book can be checked out for at most 21 days. Which inequality models the number of days d?

   (A) $d < 21$ (B) $d > 21$

   (C) $d \leq 21$ (D) $d \geq 21$

17. Which inequality does the number line below represent?

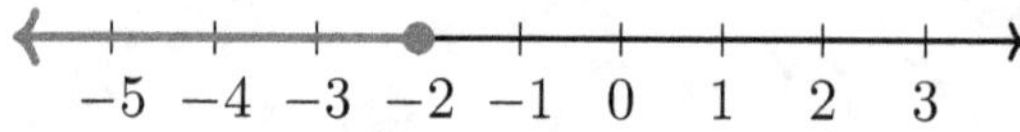

   (A) $x > -2$ (B) $x < -2$

   (C) $x \geq -2$ (D) $x \leq -2$

18. In $c = 7p$, where c is total cost and p is the number of pizzas, what is the cost of 6 pizzas?

(A) $13

(B) $36

(C) $42

(D) $67

19. A triangle has base 15 cm and height 4 cm. Sam says the area is 60 cm^2. What mistake did Sam make?

(A) He used the wrong formula entirely.

(B) He forgot to divide by 2.

(C) He added instead of multiplied.

(D) He subtracted the height from the base.

20. A toy box is 3 ft long, 2 ft wide, and 2 ft tall. What is the volume?

(A) 7 ft^3

(B) 12 ft^3

(C) 6 ft^3

(D) 24 ft^3

21. What is the distance between $(-6, -2)$ and $(-6, 5)$?

(A) 3 units

(B) 7 units

(C) 12 units

(D) 6 units

Find more at
ViewMath.com/NE-Grade6

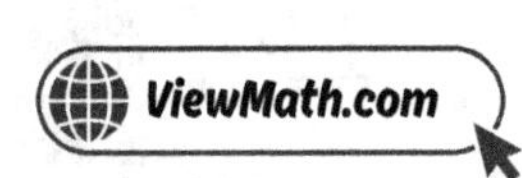

22. Find the area of the shaded triangle on the coordinate plane below.

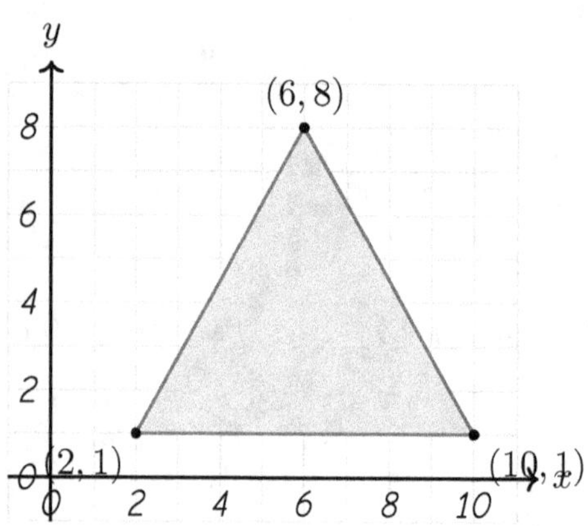

Your Answer:

23. A rectangular prism has a surface area of 94 cm². Its length is 5 cm, width is 3 cm, and height is 4 cm. Is this correct?

(A) Yes, the surface area is 94 cm².

(B) No, the surface area is 120 cm².

(C) No, the surface area is 60 cm².

(D) No, the surface area is 47 cm².

24. Which formula gives the area of a circle?

(A) $A = 2\pi r$

(B) $A = \pi d$

(C) $A = \pi r^2$

(D) $A = 2\pi r^2$

25. Data set A: $20, 21, 22, 23, 24$. Data set B: $10, 15, 22, 29, 34$. Both have a center near 22. What is different?

(A) Data set A is more spread out.

(B) Data set B is more spread out.

(C) Both have the same spread.

(D) Neither data set has a center.

Find more at
ViewMath.com/NE-Grade6

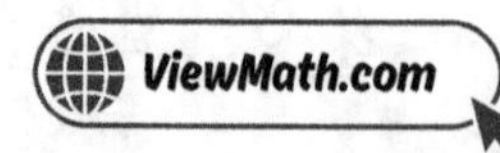

26. Data: $3, 4, 5, 6, 7, 8, 60$. *Find the mean and median. Which better describes a typical value?*

Your Answer:

27. *Which statement about MAD is true?*

(A) MAD measures the middle value of the data.

(B) MAD tells you the average distance of each value from the mean.

(C) MAD is always larger than the range.

(D) MAD equals the range divided by 2.

28. *A box plot has* min $= 10$, $Q1 = 20$, median $= 30$, $Q3 = 40$, max $= 50$. *What is the IQR?*

(A) 10

(B) 20

(C) 30

(D) 40

29. *Class A:* median $= 78$, IQR $= 6$. *Class B:* median $= 78$, IQR $= 18$. *What can you conclude?*

(A) Both classes performed identically.

(B) Class B is more consistent.

(C) Class A is more consistent because its IQR is smaller.

(D) Class B has a higher center.

30. *A bag has 5 orange balls, 3 purple balls, and 2 yellow balls. What is the probability of **not** drawing a yellow ball?*

(A) $\dfrac{2}{10}$

(B) $\dfrac{5}{10}$

(C) $\dfrac{8}{10}$

(D) $\dfrac{3}{10}$

Find more at
ViewMath.com/NE-Grade6

ViewMath.com

 # End of Practice Test 3 

Great job finishing the test!

 My Score

I got _____________ out of 30 questions right.

Check your answers in the **Answer Key** at the back of the book.

💡 Review any questions you missed. That's how we learn!

📊 Check Your Score Online!

Visit **ViewMath Academy** to enter your answers and see which topics you need to review. You can also explore lessons, take quizzes, track your scores, and save your progress!

viewmath.com/score/6.1.NE.18

Or go to viewmath.com/score and enter code: 6.1.NE.18

Practice Test 4

 *30 Questions*

✏️ Before You Start ✏️

- ✔ **Read each question carefully** before choosing your answer.
- ✔ **Show your work** on scratch paper when you need to.
- ✔ **Skip hard questions** and come back to them later.
- ✔ **Check your answers** when you're done.
- ✔ **Take your time** — there's no rush!

 **You've Got This!** 

Do your best and show what you know!

1. A store sells 3 pencils **for every** 1 eraser. Which ratio represents pencils to erasers?

   (A) $1:3$ (B) $3:1$

   (C) $3:4$ (D) $1:4$

2. A bike travels at a unit rate of 14 miles per hour. How long will it take to travel 49 miles?

   Your Answer:

3. The ratio table below has an error in one row. Which row has the error?

x	y
4	10
8	20
12	25
16	40

   (A) Row 1   (B) Row 2

   (C) Row 3   (D) Row 4

4. On a graph, Line A passes through $(1, 6)$ and Line B passes through $(1, 4)$. Both go through the origin. Which line represents a faster rate? Explain.

   Your Answer:

5. A car travels 180 miles in 3 hours. At the same rate, how far will it go in 7 hours?

   (A) 360 miles   (B) 420 miles

   (C) 540 miles   (D) 300 miles

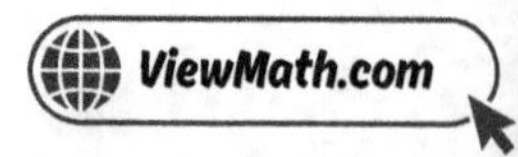

6. How many cups are in 3 quarts? (1 quart = 4 cups)

(A) 7

(B) 8

(C) 10

(D) 12

7. In the standard long-division algorithm, the four repeating steps in order are: Divide, _______, Subtract, Bring Down. What is the missing step?

(A) Estimate

(B) Multiply

(C) Check

(D) Add

8. Compute 3.6×0.4.

Your Answer:

9. A diver is at -40 feet and a bird is at 25 feet above sea level. Who is farther from sea level?

(A) The bird, because $25 > -40$

(B) They are the same distance from sea level

(C) The diver, because $|-40| > |25|$

(D) Neither, because you cannot compare positive and negative numbers

10. In which quadrant is the point $(-4, 7)$ located?

(A) Quadrant I

(B) Quadrant II

(C) Quadrant III

(D) Quadrant IV

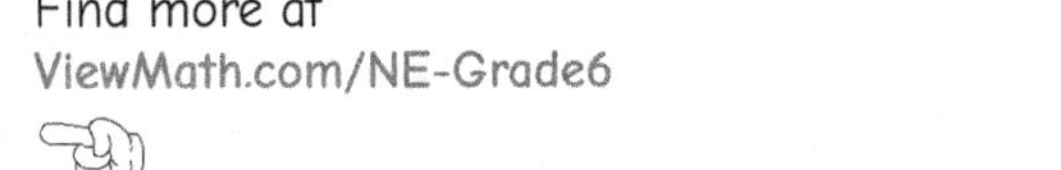

Find more at
ViewMath.com/NE-Grade6

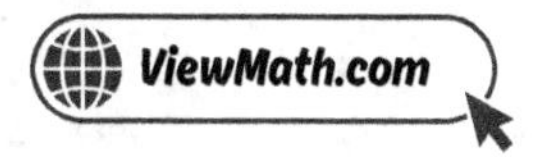

11. Which phrase matches the expression $\dfrac{n}{3} + 7$?

   (A) 7 more than a number n divided by 3

   (B) 3 divided by n plus 7

   (C) 7 times n divided by 3

   (D) n plus 7, divided by 3

12. Identify the constant(s) in the expression $3m + 7 - 2m + 4$.

   Your Answer:

13. Look at the formula and the rectangle below. What is the perimeter of the rectangle?

$$w = 3\ cm$$

$$P = 2l + 2w$$

$$l = 7\ cm$$

   (A) 10 cm

   (B) 20 cm

   (C) 21 cm

   (D) 42 cm

14. Simplify: $2(4m + 1) + 3m$

   Your Answer:

15. Solve: $k + 15 = 32$

   (A) $k = 47$

   (B) $k = 17$

   (C) $k = 27$

   (D) $k = 2$

Find more at
ViewMath.com/NE-Grade6

ViewMath.com

16. The thermometer shows the current temperature is 32°F. A weather report says "Temperatures will remain below freezing." Which inequality describes the predicted temperature t?

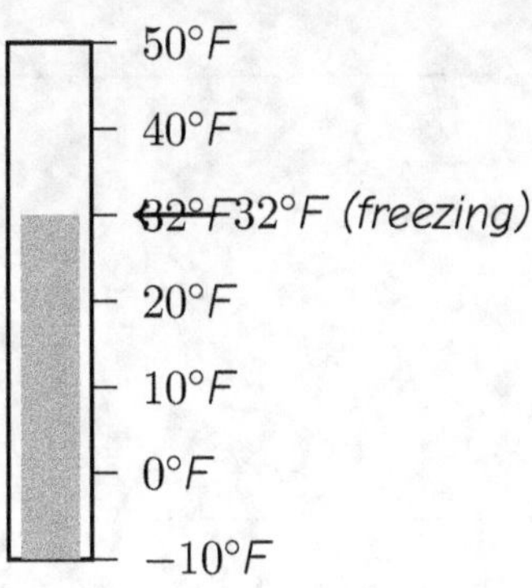

(A) $t > 32$

(B) $t \geq 32$

(C) $t < 32$

(D) $t \leq 32$

17. Which inequality has a graph where $x = -4$ is in the shaded region but $x = -2$ is NOT?

(A) $x > -3$

(B) $x < -3$

(C) $x \geq -3$

(D) $x \leq -3$

18. Using the equation $y = 3x$, complete: when $y = 21$, $x = ?$

(A) 3

(B) 7

(C) 18

(D) 63

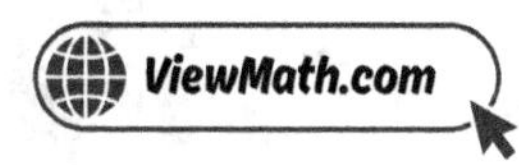

19. Two triangles both have a base of 10 cm. Triangle P has a height of 6 cm and Triangle Q has a height of 8 cm. How much greater is the area of Triangle Q?

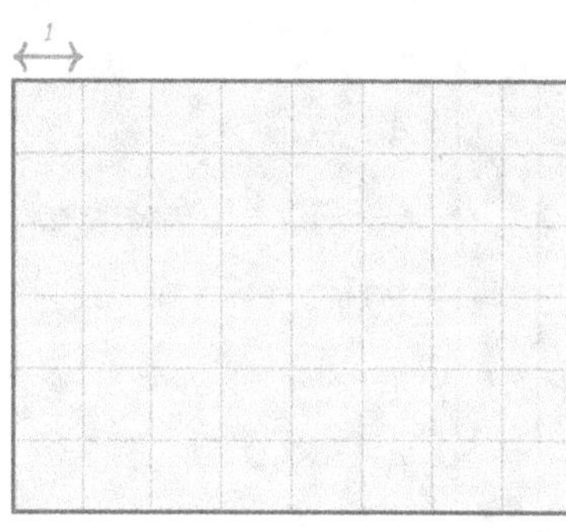

- (A) $2\ cm^2$
- (B) $10\ cm^2$
- (C) $20\ cm^2$
- (D) $40\ cm^2$

20. A rectangular prism has length $\frac{1}{2}$ m, width $\frac{1}{2}$ m, and height $\frac{1}{2}$ m. What is the volume?

- (A) $\frac{1}{8}\ m^3$
- (B) $\frac{1}{4}\ m^3$
- (C) $\frac{3}{2}\ m^3$
- (D) $\frac{1}{2}\ m^3$

21. A rectangle has vertices $(1,2)$, $(7,2)$, $(7,-4)$, and $(1,-4)$. What is the length of the longer side?

- (A) 5 units
- (B) 6 units
- (C) 7 units
- (D) 8 units

22. A rectangle has vertices $(-1,3)$, $(5,3)$, $(5,-2)$, and $(-1,-2)$. What is the area?

- (A) 25 square units
- (B) 30 square units
- (C) 22 square units
- (D) 35 square units

Find more at
ViewMath.com/NE-Grade6

23. Which net below could fold into a cube?

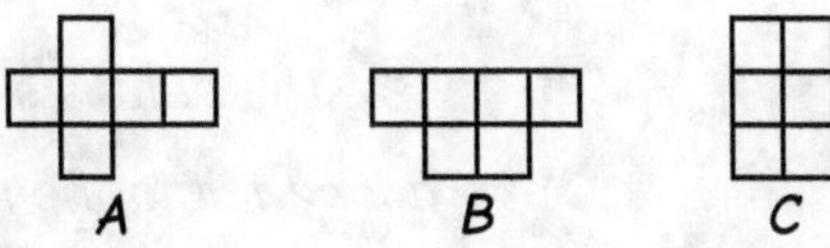

(A) Net A only

(B) Net B only

(C) Net C only

(D) Nets A and B

24. A sprinkler waters a circular area with a radius of 8 meters. What is the total area that the sprinkler covers? Use $\pi \approx 3.14$.

Your Answer:

25. Data set: $22, 24, 25, 25, 26, 27, 45$. Which value is most likely an outlier?

(A) 22

(B) 25

(C) 27

(D) 45

26. Data: $11, 13, 15, 17, 19, 21, 23$. What are the mean and median?

(A) Mean = 17, Median = 17

(B) Mean = 17, Median = 15

(C) Mean = 15, Median = 17

(D) Mean = 19, Median = 17

27. If every value in a data set is increased by 5, what happens to the range?

(A) It increases by 5.

(B) It decreases by 5.

(C) It stays the same.

(D) It doubles.

Find more at
ViewMath.com/NE-Grade6

ViewMath.com

28. A box plot has a very long right whisker and a short left whisker. What does this suggest?

(A) The data is symmetric.

(B) The data is skewed right (has high outliers or a long tail to the right).

(C) The data is skewed left.

(D) All values are the same.

29. The table below shows statistics for two math classes on a unit test.

Statistic	Period 1	Period 3
Mean	76	82
Median	78	81
Range	40	18
IQR	16	8

Write a paragraph comparing the two classes. Use specific numbers to support your claims.

Your Answer:

30. A standard number cube is rolled. What is the probability of rolling a 5, expressed as a decimal rounded to the nearest hundredth?

(A) 0.50

(B) 0.20

(C) 0.17

(D) 0.83

 End of Practice Test 4 

Great job finishing the test!

☑ My Score

I got _____________ out of 30 questions right.

*Check your answers in the **Answer Key** at the back of the book.*

💡 *Review any questions you missed. That's how we learn!*

📊 Check Your Score Online!

Visit **ViewMath Academy** to enter your answers and see which topics you need to review. You can also explore lessons, take quizzes, track your scores, and save your progress!

viewmath.com/score/6.1.NE.19

Or go to viewmath.com/score and enter code: 6.1.NE.19

Practice Test 5

☑ 30 Questions

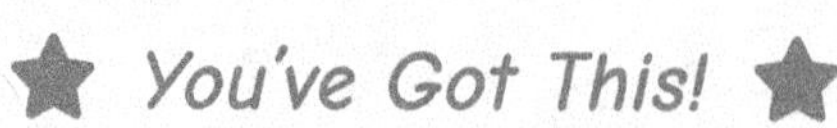

✏ Before You Start ✏

✔ **Read each question carefully** before choosing your answer.

✔ **Show your work** on scratch paper when you need to.

✔ **Skip hard questions** and come back to them later.

✔ **Check your answers** when you're done.

✔ **Take your time** — there's no rush!

★ You've Got This! ★

Do your best and show what you know!

1. A class votes on two activities: hiking and swimming. The ratio of votes for hiking to swimming is 3 : 2. There are 25 votes total. How many voted for swimming?

Your Answer:

2. A car travels 240 miles using 8 gallons of gas. What is the unit rate in miles per gallon?

(A) 8

(B) 32

(C) 30

(D) 248

3. It takes 6 cups of flour to make 4 loaves of bread. How many cups of flour for 10 loaves?

(A) 12

(B) 10

(C) 15

(D) 20

Find more at
ViewMath.com/NE-Grade6

4. The table and partial graph below show a ratio relationship.

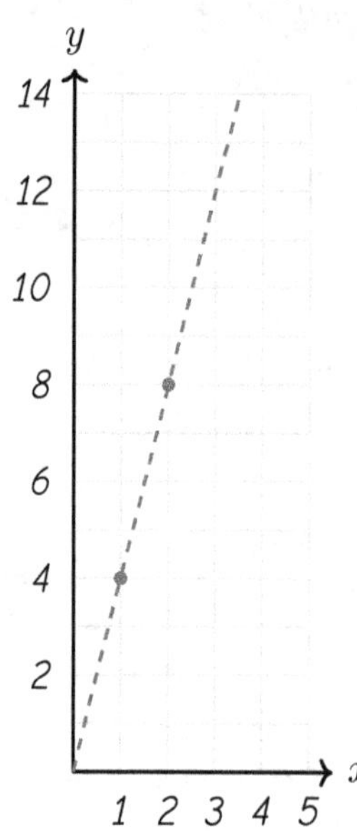

x	y
1	4
2	8
3	?

What is the missing value of y when $x = 3$?

(A) 10

(B) 12

(C) 14

(D) 16

5. A map scale says 1 cm = 5 km. Two cities are 8 cm apart on the map. What is the real distance?

(A) 13 km

(B) 5 km

(C) 40 km

(D) 80 km

6. Convert 2 hours, 30 minutes to minutes, then to seconds.

Your Answer:

Find more at
ViewMath.com/NE-Grade6

ViewMath.com

7. What is $1{,}575 \div 5$?

(A) 305

(B) 3,150

(C) 315

(D) 351

8. When you multiply 0.6×0.7, how many decimal places should the product have?

(A) 0

(B) 1

(C) 2

(D) 3

9. Which statement is **true**?

(A) $|-10| = -10$

(B) $|-10| = 10$

(C) $|-10| = 0$

(D) $|-10| = -(-(-10))$

10. Starting at the origin, you move 4 units to the left and 3 units up, then 7 units to the right and 5 units down. What point do you end at?

Your Answer:

11. Write an expression for: "9 fewer than a number z."

Your Answer:

12. In the expression $6(n+2)$, what are the two factors?

(A) 6 and n

(B) 6 and 2

(C) n and 2

(D) 6 and $(n+2)$

Find more at
ViewMath.com/NE-Grade6

ViewMath.com

13. Evaluate $m^2 + 2m + 1$ when $m = 4$.

Your Answer:

14. Simplify: $4(3y - 2)$

(A) $12y - 2$

(B) $12y - 8$

(C) $7y - 2$

(D) $7y - 6$

15. Solve: $8w = 72$

(A) $w = 8$

(B) $w = 64$

(C) $w = 9$

(D) $w = 80$

16. Is $n = 4$ a solution to $n > 4$? Is $n = 4$ a solution to $n \geq 4$? Explain both.

Your Answer:

17. A student graphs $x > 2$ with a closed circle at 2 and shading to the right. What is the error?

(A) Should shade to the left

(B) Should use an open circle at 2

(C) Should use a circle at 0 instead

(D) There is no error

18. A plant grows 3 cm each week. Which equation relates the height h (in cm) to the number of weeks w?

(A) $h = w + 3$

(B) $h = 3w$

(C) $w = 3h$

(D) $h = w \div 3$

Find more at
ViewMath.com/NE-Grade6

ViewMath.com

19. A triangular pennant has a base of 5 in and a height of 12 in. What is its area?

(A) $60\ in^2$ (B) $17\ in^2$

(C) $30\ in^2$ (D) $34\ in^2$

20. If you double the height of a rectangular prism but keep the length and width the same, what happens to the volume?

(A) The volume stays the same.   (B) The volume doubles.

(C) The volume triples.   (D) The volume quadruples.

21. A square on the coordinate plane has vertices $(0,0)$, $(0,5)$, $(5,5)$, and $(5,0)$. What is the perimeter?

Your Answer:

22. A right triangle has vertices $(2,1)$, $(2,7)$, and $(8,1)$. What is the area?

(A) 36 square units   (B) 18 square units

(C) 12 square units   (D) 24 square units

23. A box is 5 in long, 5 in wide, and 10 in tall. What is the surface area?

(A) $250\ in^2$ (B) $200\ in^2$

(C) $300\ in^2$ (D) $150\ in^2$

24. What does the number π (pi) represent?

(A) The radius divided by the diameter   (B) The ratio of a circle's circumference to its diameter

(C) The area of any circle with radius 1   (D) The diameter divided by the circumference

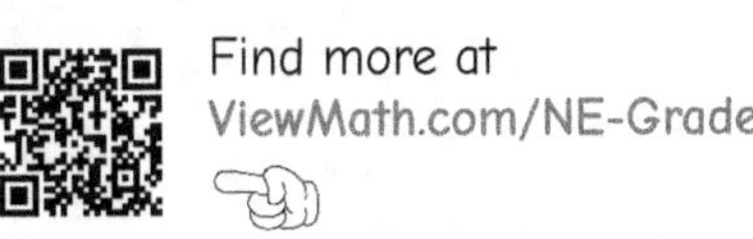
Find more at
ViewMath.com/NE-Grade6

25. Data: $5, 5, 6, 6, 6, 7, 7, 7, 7, 8, 8.$ Where do the values cluster?

(A) Around 5

(B) Around 6–7

(C) Around 8

(D) The data does not cluster.

26. The mean of five numbers is 20. What is the sum of the five numbers?

(A) 4

(B) 25

(C) 80

(D) 100

27. Data: $6, 8, 10, 12, 14.$ The mean is 10. Find the MAD.

Your Answer:

28. A box plot has the median line closer to Q1 than to Q3. What does this mean?

(A) The data is perfectly symmetric.

(B) The upper half of the data is more spread out than the lower half.

(C) The lower half is more spread out.

(D) The data has no outliers.

Find more at
ViewMath.com/NE-Grade6

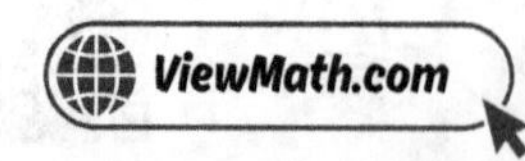

29. The two dot plots below show the daily tips (in dollars) earned by two servers at a restaurant.

Server A

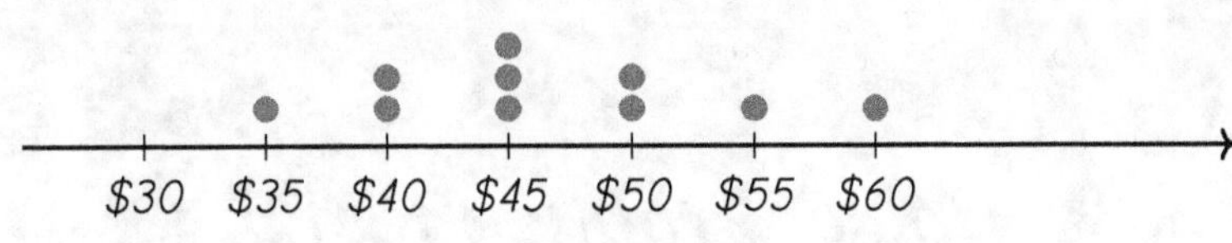

Server B

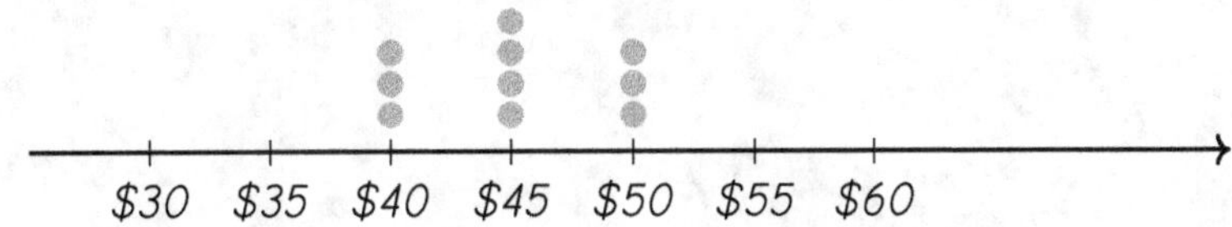

Find the median for each server. Which server earns more on a typical day? Which server's tips are more consistent?

Your Answer:

30. Event A has a probability of $\frac{2}{5}$ and Event B has a probability of $\frac{3}{4}$. Which statement is true?

(A) Event A is more likely than Event B

(B) Event B is more likely than Event A

(C) Both events are equally likely

(D) Neither event can happen

 # End of Practice Test 5 

Great job finishing the test!

 My Score

I got ___________ out of 30 questions right.

Check your answers in the **Answer Key** at the back of the book.

💡 Review any questions you missed. That's how we learn!

📊 Check Your Score Online!

Visit **ViewMath Academy** to enter your answers and see which topics
you need to review. You can also explore lessons, take quizzes, track
your scores, and save your progress!

viewmath.com/score/6.1.NE.20

Or go to *viewmath.com/score* and enter code: 6.1.NE.20

Practice Test 6

 30 Questions

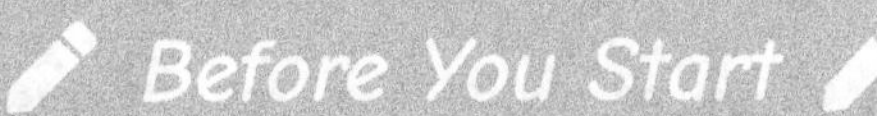

✏ Before You Start ✏

- ✔ **Read each question carefully** before choosing your answer.
- ✔ **Show your work** on scratch paper when you need to.
- ✔ **Skip hard questions** and come back to them later.
- ✔ **Check your answers** when you're done.
- ✔ **Take your time** — there's no rush!

⭐ You've Got This! ⭐

Do your best and show what you know!

1. A smoothie recipe uses 3 bananas **for every** 4 cups of yogurt. Write the ratio of bananas to yogurt. Then write the ratio of yogurt to bananas.

Your Answer:

2. Brand X sells 5 notebooks for $8.75. Brand Y sells 3 notebooks for $4.50. Which is the better deal?

(A) Brand X at $1.75 each

(B) Brand Y at $1.50 each

(C) Brand X at $1.50 each

(D) Brand Y at $1.75 each

3. A teacher hands out 3 pencils for every 2 students. There are 16 students. How many pencils does she need?

(A) 18

(B) 24

(C) 20

(D) 32

4. A car travels at 30 miles per hour. Which set of points represents this ratio?

(A) $(1, 30), (2, 60), (3, 90)$

(B) $(30, 1), (60, 2), (90, 3)$

(C) $(1, 30), (2, 50), (3, 90)$

(D) $(1, 3), (2, 6), (3, 9)$

5. Which strategy is best for solving "how much for one?" problems?

(A) Tape diagram

(B) Unit rate

(C) Ratio table

(D) Equation

6. Convert 7,000 grams to kilograms.

(A) 0.7 kg

(B) 7 kg

(C) 70 kg

(D) 700 kg

Find more at
ViewMath.com/NE-Grade6

7. A baker makes 2,016 cookies and packs 14 cookies per bag. How many bags can she fill?

   (A) 142   (B) 144

   (C) 148   (D) 104

8. What is $9.36 \div 1.2$?

   (A) 78   (B) 0.78

   (C) 7.8   (D) 7.08

9. What is the opposite of -3?

   (A) -3 (B) 0

   (C) $\dfrac{1}{3}$ (D) 3

10. Starting at the origin, how do you plot the point $(-3, 4)$?

   (A) Move 3 units right and 4 units up   (B) Move 3 units left and 4 units down

   (C) Move 3 units left and 4 units up   (D) Move 4 units left and 3 units up

11. Which expression represents "6 more than half of a number n"?

   (A) $6 + 2n$ (B) $\dfrac{n + 6}{2}$

   (C) $\dfrac{n}{2} + 6$ (D) $n + 3$

12. Look at the expression map below. Which label correctly identifies Part C?

$$\underbrace{5}_{A} \underbrace{x}_{B} + \underbrace{3}_{C}$$

(A) A variable

(B) A coefficient

(C) A constant

(D) A factor

13. Evaluate $\dfrac{x^2}{4}$ when $x = 8$.

Your Answer

14. Simplify: $3(x + 4) + 2x$

(A) $5x + 4$

(B) $5x + 12$

(C) $6x + 12$

(D) $3x + 6$

15. Solve: $\dfrac{t}{6} = 5$

(A) $t = 11$

(B) $t = 1$

(C) $t = 30$

(D) $t = 56$

16. Which inequality represents "you need more than \$25 to buy the video game"?

(A) $d \leq 25$

(B) $d < 25$

(C) $d > 25$

(D) $d \geq 25$

17. *Write the inequality that matches this description: closed circle at −4, shade to the left.*

Your Answer:

18. *A school bus holds 50 students. How many buses b are needed for s students? Write an equation.*

Your Answer:

19. *A triangular garden has a base of 12 ft and a height of 9 ft. How many square feet of soil are needed to cover it?*

(A) $108\ ft^2$

(B) $54\ ft^2$

(C) $21\ ft^2$

(D) $42\ ft^2$

20. *A rectangular prism is made entirely of 1-cm cubes. It is 4 cubes long, 3 cubes wide, and 5 cubes tall. What is the volume?*

Your Answer:

21. *Which pair of points forms a vertical segment?*

(A) $(3, 2)$ and $(7, 2)$

(B) $(5, -1)$ and $(5, 4)$

(C) $(1, 3)$ and $(4, 6)$

(D) $(0, 0)$ and $(3, 0)$

22. *A rectangle has vertices $(2, -1)$, $(2, 5)$, $(9, 5)$, and $(9, -1)$. What is the area?*

Your Answer:

23. A rectangular prism has length 8 cm, width 5 cm, and height 2 cm. Leo says the surface area is 80 cm^2. What did Leo calculate instead?

(A) The volume

(B) The perimeter

(C) Half the surface area

(D) The area of just one face

24. A circle has a diameter of 12 m. What is its area? Use $\pi \approx 3.14$.

(A) 37.68 m^2

(B) 113.04 m^2

(C) 226.08 m^2

(D) 452.16 m^2

25. Data: $15, 16, 16, 17, 17, 17, 18, 100$. How does the outlier 100 affect the center?

(A) It pulls the center toward 100, making it higher than typical values.

(B) It has no effect on the center.

(C) It makes the center equal to 100.

(D) It pulls the center lower.

26. Data: $8, 8, 9, 10, 10, 11, 50$. Which is the better measure of center?

(A) Mean, because it uses all the data.

(B) Median, because the outlier 50 pulls the mean too high.

(C) Neither is useful.

(D) Both are equally good.

27. Data: $1, 2, 3, 4, 5, 6, 100$. What is the range?

(A) 5

(B) 6

(C) 94

(D) 99

Find more at
ViewMath.com/NE-Grade6

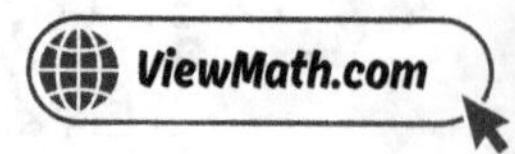

28. Data (in order): $3, 5, 7, 9, 11, 13, 15$. What is Q1?

(A) 3

(B) 5

(C) 7

(D) 9

29. When should you use the median and IQR instead of the mean and MAD?

(A) When the data is perfectly symmetric.

(B) When the data has outliers or is skewed.

(C) When every value is the same.

(D) When the data set is very small.

30. Which of the following best describes the range of all probabilities?

(A) From -1 to 1

(B) From 0 to 1

(C) From 0 to 100

(D) From 1 to 10

 # End of Practice Test 6 

Great job finishing the test!

☑ My Score

I got __________ out of 30 questions right.

*Check your answers in the **Answer Key** at the back of the book.*

💡 *Review any questions you missed. That's how we learn!*

📊 Check Your Score Online!

Visit **ViewMath Academy** to enter your answers and see which topics you need to review. You can also explore lessons, take quizzes, track your scores, and save your progress!

viewmath.com/score/6.1.NE.21

*Or go to **viewmath.com/score** and enter code: 6.1.NE.21*

Practice Test 7

 30 Questions

 Before You Start

- ✔ **Read each question carefully** before choosing your answer.
- ✔ **Show your work** on scratch paper when you need to.
- ✔ **Skip hard questions** and come back to them later.
- ✔ **Check your answers** when you're done.
- ✔ **Take your time** — there's no rush!

⭐ You've Got This! ⭐

Do your best and show what you know!

1. A lemonade stand sells 7 cups of lemonade **for each** 2 cups of iced tea. If they sold 21 cups of lemonade, how many cups of iced tea did they sell?

Your Answer:

2. A garden hose fills a 150-gallon tank in 5 hours. What is the unit rate?

(A) 25 gallons per hour

(B) 30 gallons per hour

(C) 35 gallons per hour

(D) 750 gallons per hour

3. Complete the ratio table for the ratio 5 : 8.

5	8
10	?
?	32

Your Answer:

4. A line passes through the origin and the point $(5, 15)$. What is the ratio x to y?

(A) 1 : 5

(B) 5 : 1

(C) 1 : 3

(D) 3 : 1

5. Two families split a $350 dinner bill in a 3 : 4 ratio. How much does each family pay?

Your Answer:

Find more at
ViewMath.com/NE-Grade6

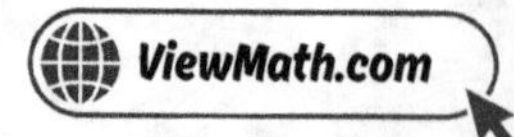

6. How many inches are in 5 feet?

(A) 50   (B) 55

(C) 60   (D) 72

7. Compute $9,072 \div 42$.

Your Answer:

8. Sam buys 3 folders that each cost \$2.75. How much does he spend in all?

(A) \$8.75   (B) \$8.25

(C) \$6.25   (D) \$5.75

9. What is the opposite of 5?

(A) 5   (B) -5

(C) 0   (D) $\dfrac{1}{5}$

10. Where is the point $(-6, 0)$ located on the coordinate plane?

(A) On the y-axis   (B) On the x-axis

(C) In Quadrant II   (D) In Quadrant III

11. Write a word phrase for the expression $5x + 3$.

Your Answer:

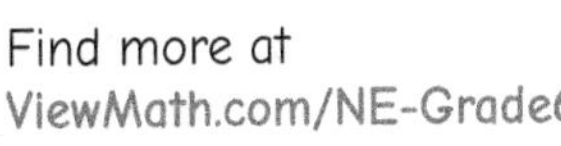

12. Which expression has exactly 4 terms?

(A) $2x + 3y$

(B) $a + b + c$

(C) $5m - 2n + 7 + m$

(D) $4p$

13. The perimeter of a rectangle is $P = 2l + 2w$. What is P when $l = 9$ and $w = 4$?

(A) 13

(B) 22

(C) 26

(D) 36

14. Simplify: $5p + 3 + 2p + 4p - 1$

(A) $11p + 2$

(B) $7p + 2$

(C) $11p + 4$

(D) $12p$

15. Solve: $m + 17 = 30$

Your Answer:

16. A bag can hold at most 50 pounds. Which inequality represents the allowed weight w?

(A) $w > 50$

(B) $w < 50$

(C) $w \leq 50$

(D) $w \geq 50$

17. When graphing $x > 4$ on a number line, what kind of circle goes at 4?

(A) Closed circle (filled in)

(B) Open circle (not filled in)

(C) No circle — just an arrow

(D) A square

Find more at
ViewMath.com/NE-Grade6

18. On a graph of $y = 4x$, which ordered pair is on the line?

(A) $(2, 6)$

(B) $(3, 12)$

(C) $(4, 12)$

(D) $(5, 25)$

19. Which of the following is true about the height of a triangle?

(A) The height is always a side of the triangle.

(B) The height must be perpendicular to the base.

(C) The height is always the longest measurement.

(D) The height must equal the base.

20. A box is 2.5 cm long, 4 cm wide, and 6 cm tall. What is the volume?

Your Answer:

21. A triangle on the coordinate plane has vertices $(0, 0)$, $(8, 0)$, and $(4, 6)$. What is the length of the base along the x-axis?

(A) 4 units

(B) 6 units

(C) 8 units

(D) 10 units

22. A square has vertices $(-2, -2)$, $(4, -2)$, $(4, 4)$, and $(-2, 4)$. What is the area?

(A) 24 square units

(B) 36 square units

(C) 12 square units

(D) 16 square units

23. Find the surface area of a rectangular prism with length 7 m, width 5 m, and height 3 m.

Your Answer:

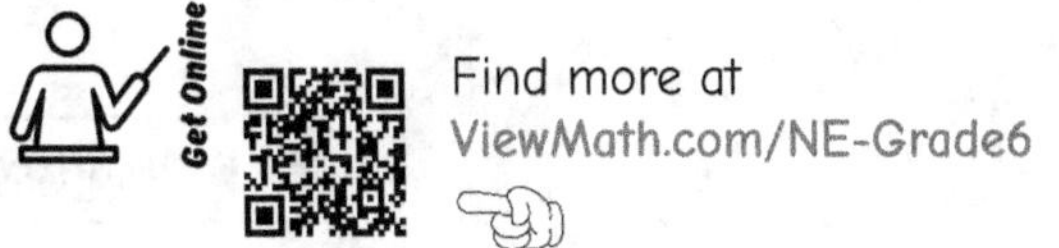

Find more at
ViewMath.com/NE-Grade6

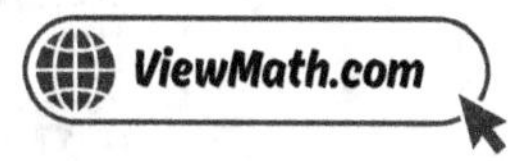

24. Two circles are shown below. Circle P has a radius of 4 in and Circle Q has a radius of 8 in. The area of Circle Q is how many times the area of Circle P?

(A) 2 times

(B) 3 times

(C) 4 times

(D) 8 times

25. Two students collected data. Student A: range = 5. Student B: range = 40. Whose data is more consistent?

Your Answer

26. The mean of 6 quiz scores is 15. If one score of 9 is removed, what is the mean of the remaining 5 scores?

(A) 14.2

(B) 15

(C) 16.2

(D) 18

27. The table below shows the hourly wages of 7 employees at a store.

Employee	A	B	C	D	E	F	G
Wage ($)	9	10	10	12	14	15	30

Find the range and IQR. Explain which measure better describes the typical spread of wages.

Your Answer:

Find more at
ViewMath.com/NE-Grade6

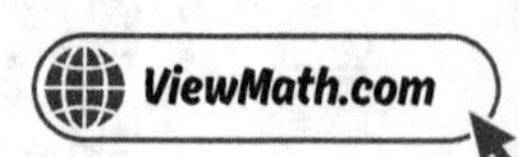

28. Which five values make up the five-number summary?

(A) Mean, median, mode, range, IQR

(B) Minimum, Q1, median, Q3, maximum

(C) Q1, Q2, Q3, Q4, Q5

(D) Mean, Q1, Q3, range, MAD

29. Data: $10, 12, 14, 16, 18, 20, 22$. The mean is 16. A student says the data is "very spread out." Is this correct?

(A) Yes, the range is 12.

(B) No, the range (12) is moderate and the values increase by a steady 2.

(C) Yes, because there are 7 values.

(D) No, because the mean equals the median.

30. A bag contains 8 marbles: 2 are yellow. What is the probability of drawing a yellow marble, written as a decimal?

(A) 0.8

(B) 0.28

(C) 0.2

(D) 0.25

 End of Practice Test 7 

Great job finishing the test!

 My Score

I got _____________ out of 30 questions right.

*Check your answers in the **Answer Key** at the back of the book.*

 Review any questions you missed. That's how we learn!

Check Your Score Online!

Visit **ViewMath Academy** to enter your answers and see which topics you need to review. You can also explore lessons, take quizzes, track your scores, and save your progress!

viewmath.com/score/6.1.NE.22

Or go to viewmath.com/score and enter code: 6.1.NE.22

Practice Test 8

 30 Questions

✏️ Before You Start ✏️

- ✓ **Read each question carefully** before choosing your answer.
- ✓ **Show your work** on scratch paper when you need to.
- ✓ **Skip hard questions** and come back to them later.
- ✓ **Check your answers** when you're done.
- ✓ **Take your time** — there's no rush!

⭐ **You've Got This!** ⭐

Do your best and show what you know!

1. The ratio of cats to dogs at a pet store is $3 : 4$. Each part in the tape diagram represents 2 animals. How many dogs are there?

(A) 6

(B) 8

(C) 3

(D) 4

2. The graph shows the number of laps a swimmer completes over time.

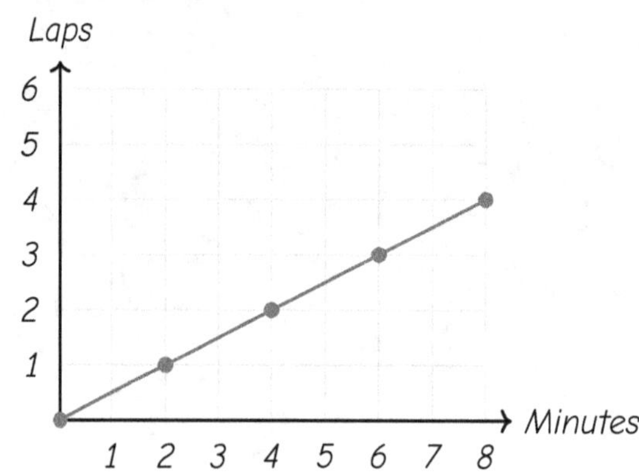

Part A: What is the swimmer's unit rate in laps per minute?

Part B: How many laps will the swimmer complete in 14 minutes?

Your Answer:

3. A smoothie recipe uses 2 cups of strawberries for every 3 cups of yogurt. How many cups of yogurt are needed for 10 cups of strawberries?

Your Answer:

4. A ratio graph passes through $(0,0)$ and $(4,7)$. What is y when $x = 8$?

(A) 11

(B) 14

(C) 15

(D) 12

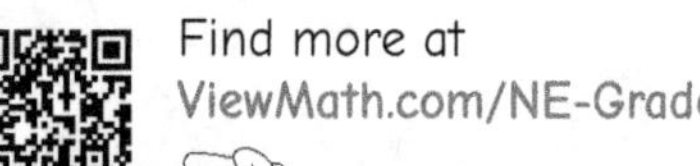
Find more at
ViewMath.com/NE-Grade6

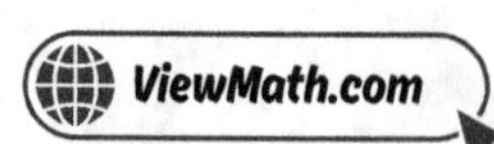

5. A printing company prints 250 flyers in 10 minutes. How long will it take to print 1,000 flyers?

Your Answer:

6. The bar model below compares different length units.

1 yard		
1 ft	1 ft	1 ft

If a rope is 7 yards long, how many feet is it?

(A) 10

(B) 14

(C) 21

(D) 28

7. Which multiplication equation can be used to check that $4{,}752 \div 12 = 396$?

(A) $396 \times 12 = 4{,}752$

(B) $396 + 12 = 408$

(C) $4{,}752 \times 12 = 57{,}024$

(D) $396 \div 12 = 33$

8. What is 0.12×0.5?

(A) 0.6

(B) 0.06

(C) 0.006

(D) 6.0

9. What is the opposite of 0?

(A) 1

(B) -1

(C) 0

(D) There is no opposite of 0

10. **Part A:** In which quadrant is the point $(-8, 1)$ located?

Part B: If $(-8, 1)$ is reflected across the y-axis, what are the coordinates of its image and in which quadrant does it lie?

Part C: If the original point $(-8, 1)$ is instead reflected across the x-axis, what are the coordinates of its image and in which quadrant does it lie?

Your Answer:

11. The table below shows the relationship between a word phrase and an expression. Which expression belongs in the blank?

Word Phrase	Expression
A number plus 4	$n + 4$
3 times a number	$3n$
8 less than a number	?
A number divided by 2	$n \div 2$

(A) $8 - n$

(B) $n - 8$

(C) $8n$

(D) $n + 8$

12. What is the coefficient of d in the expression $15 - d + 3$?

(A) 0

(B) 1

(C) -1

(D) 15

13. A parking garage charges \$5 plus \$3 per hour. How much does it cost to park for 6 hours? Use the expression $5 + 3h$.

Your Answer:

Find more at
ViewMath.com/NE-Grade6

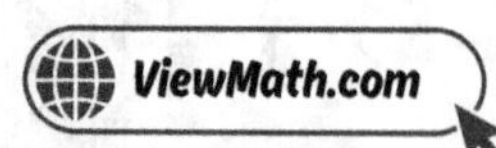

14. *The rectangle below has the dimensions shown. Write a simplified expression for its perimeter.*

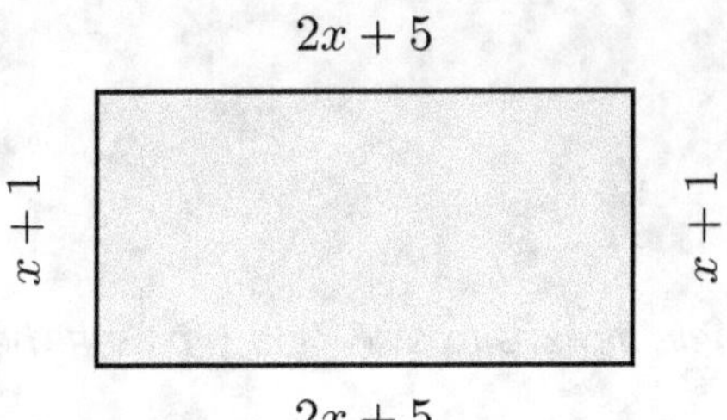

Your Answer:

15. *Write and solve an equation: A number decreased by 19 is 31.*

Your Answer:

16. *A sign says "No more than 4 guests per room." Write an inequality for the number of guests g.*

Your Answer:

17. *Is $x = 3$ a solution to $x \leq 3$? Is $x = 3$ a solution to $x < 3$? Explain.*

Your Answer:

18. *In the equation $y = x + 4$, which statement is true?*

(A) *x is the dependent variable*

(B) *y is always 4*

(C) *x is the independent variable*

(D) *y decreases as x increases*

Find more at
ViewMath.com/NE-Grade6

ViewMath.com

19. A triangular flower bed has a base of 4.5 m and a height of 6 m. What is its area?

Your Answer:

20. Two boxes are shown below with their dimensions. Which box has the greater volume?

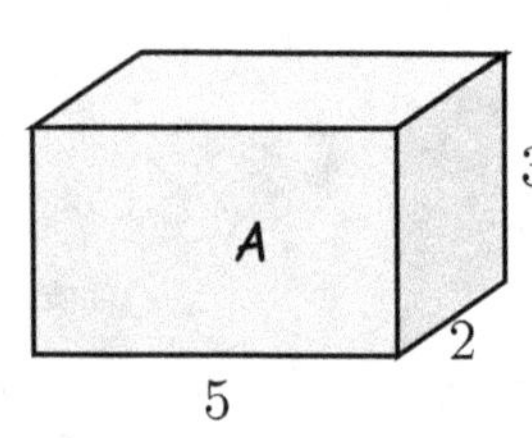
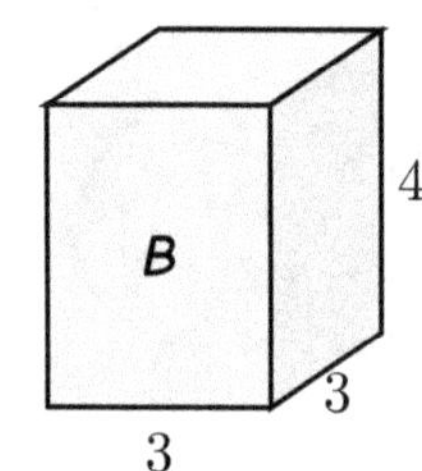

(A) Box A ($V = 30$)

(B) Box B ($V = 36$)

(C) They have the same volume.

(D) Box B ($V = 30$)

21. Three vertices of a rectangle are $(2, 3)$, $(2, -5)$, and $(9, 3)$. What is the fourth vertex?

Your Answer:

22. A right triangle has vertices $(0, 0)$, $(8, 0)$, and $(0, 6)$. What is the area?

(A) 48 square units

(B) 24 square units

(C) 14 square units

(D) 28 square units

23. What is a net?

(A) A 3D shape made of cubes

(B) A flat pattern that folds into a 3D shape

(C) The volume of a rectangular prism

(D) A grid used to measure area

Find more at
ViewMath.com/NE-Grade6

ViewMath.com

24. A circle has a diameter of 18 inches. What is its area? Use $\pi \approx 3.14$.

Your Answer:

25. Describe the shape of the data set: $1, 2, 3, 3, 4, 4, 4, 5, 5, 5, 5$.

Your Answer:

26. Data: $2, 3, 4, 5, 100$. The mean is 22.8. Why is the mean much higher than most of the values?

(A) There are only 5 values.

(B) The median is too low.

(C) The outlier 100 pulls the mean up.

(D) The values are all positive.

27. Data: $50, 55, 60, 65, 70$. The mean is 60. Find the MAD.

Your Answer:

28. Two box plots have the same median. Class A has IQR $= 8$ and Class B has IQR $= 24$. What can you conclude?

(A) Class A and Class B performed equally.

(B) Class A's scores are more spread out.

(C) Class B's middle 50% of scores are more spread out.

(D) The range of Class A is larger.

29. Class A: median $= 90$, IQR $= 6$. Class B: median $= 82$, IQR $= 18$. Compare the two classes.

Your Answer:

Find more at
ViewMath.com/NE-Grade6

30. A standard number cube is rolled. What is the probability of rolling a number less than or equal to 3?

(A) $\dfrac{1}{3}$

(B) $\dfrac{3}{6}$

(C) $\dfrac{2}{6}$

(D) $\dfrac{4}{6}$

 # End of Practice Test 8 

Great job finishing the test!

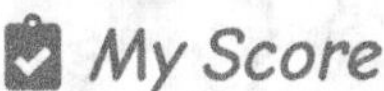 My Score

I got _____________ out of 30 questions right.

Check your answers in the **Answer Key** at the back of the book.

Review any questions you missed. That's how we learn!

📊 Check Your Score Online!

Visit **ViewMath Academy** to enter your answers and see which topics you need to review. You can also explore lessons, take quizzes, track your scores, and save your progress!

viewmath.com/score/6.1.NE.23

Or go to viewmath.com/score and enter code: 6.1.NE.23

Practice Test 9

📋 30 Questions

✏️ Before You Start ✏️

- ✔ **Read each question carefully** before choosing your answer.
- ✔ **Show your work** on scratch paper when you need to.
- ✔ **Skip hard questions** and come back to them later.
- ✔ **Check your answers** when you're done.
- ✔ **Take your time** — there's no rush!

⭐ You've Got This! ⭐

Do your best and show what you know!

1. The ratio of red to blue to green beads is $1 : 3 : 2$. There are 18 beads total. How many blue beads are there?

(A) 3

(B) 6

(C) 9

(D) 12

2. A printer prints 120 pages in 4 minutes. What is the unit rate?

(A) 4 pages per minute

(B) 30 pages per minute

(C) 120 pages per minute

(D) 480 pages per minute

3. Look at this ratio table. What is the missing value?

Apples	Oranges
3	5
6	?

(A) 8

(B) 10

(C) 15

(D) 11

4. Does the point $(3, 9)$ belong on the graph of the ratio $1 : 3$?

(A) No, because $3 + 9 \neq 1 + 3$.

(B) Yes, because $1 : 3 = 3 : 9$.

(C) No, because $3 \times 3 \neq 1$.

(D) Yes, because $3 + 9 = 12$.

5. A batch of trail mix uses 4 cups of granola for every 3 cups of raisins. If you have 12 cups of raisins, how many cups of granola do you need?

(A) 9

(B) 12

(C) 16

(D) 15

Find more at
ViewMath.com/NE-Grade6

ViewMath.com

6. A recipe needs 2.5 liters of water. How many milliliters is that?

   (A) 25

   (B) 250

   (C) 2,500

   (D) 25,000

7. What is $2,688 \div 16$?

   (A) 158

   (B) 168

   (C) 178

   (D) 1,068

8. The table shows the distances a cyclist rode each day.

Day	Distance (km)
Monday	8.76
Tuesday	12.30
Wednesday	6.48

Part A: What is the total distance the cyclist rode over the three days?

Part B: If the cyclist wants to divide the total distance equally over the 3 days, how many kilometers per day is that?

Your Answer:

9. A number plus its opposite always equals which value?

   (A) 1

   (B) The number itself

   (C) −1

   (D) 0

Find more at
ViewMath.com/NE-Grade6

ViewMath.com

10. A triangle is drawn on the coordinate plane below with vertices at A, B, and C.

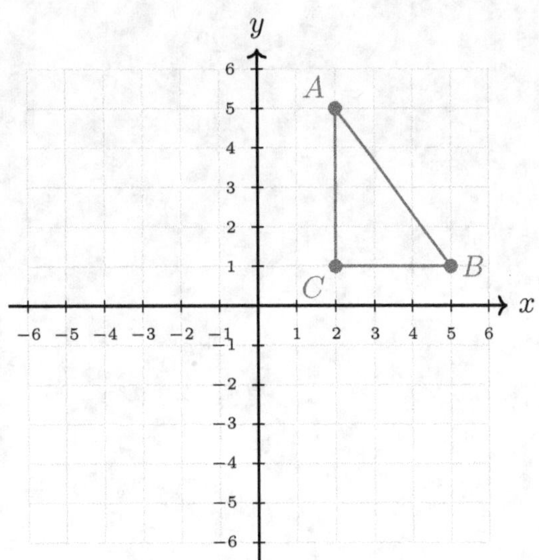

Part A: Write the coordinates of each vertex: A, B, and C.

Part B: If the triangle is reflected across the y-axis, write the coordinates of the new vertices A′, B′, and C′.

Part C: In which quadrant will the reflected triangle be located?

Your Answer:

11. Which expression represents "the sum of 9 and the product of 4 and t"?

(A) $9 + 4 + t$

(B) $9 + 4t$

(C) $9(4 + t)$

(D) $9 \times 4t$

12. How many terms are in the expression $4x + 9 - 2y$?

(A) 2

(B) 3

(C) 4

(D) 5

13. Evaluate $6x - x^2$ when $x = 4$.

(A) 8

(B) 20

(C) -8

(D) 40

14. Simplify: $7n + 4n$

(A) $11n$

(B) $28n$

(C) $11n^2$

(D) $74n$

15. Which inverse operation would you use to solve $\dfrac{n}{3} = 12$?

(A) Divide both sides by 3

(B) Subtract 3 from both sides

(C) Add 3 to both sides

(D) Multiply both sides by 3

16. Which of the following is a solution to $n \leq 8$?

(A) $n = 9$

(B) $n = 8.5$

(C) $n = 8$

(D) $n = 10$

17. Which value is NOT in the solution set of $x < 10$?

(A) $x = 0$

(B) $x = 9.9$

(C) $x = -5$

(D) $x = 10$

18. Using $d = 60t$, what is the distance after 3 hours?

(A) 20 miles

(B) 63 miles

(C) 180 miles

(D) 360 miles

Find more at
ViewMath.com/NE-Grade6

ViewMath.com

19. A triangle has base 20 in and height 7 in. What is the area?

  (A) 140 in^2 (B) 27 in^2

  (C) 70 in^2 (D) 54 in^2

20. A box has a volume of 72 in^3. The base is 6 in by 4 in. What is the height?

  (A) 2 in   (B) 3 in

  (C) 6 in   (D) 12 in

21. A square has one vertex at $(-2, -2)$ and side length 5. The sides are horizontal and vertical. Which could be the opposite vertex?

  (A) $(3, 3)$ (B) $(3, -7)$

  (C) $(5, 5)$ (D) $(2, 2)$

22. A rectangle has vertices $(0, 0)$, $(6, 0)$, $(6, 4)$, and $(0, 4)$. What is the area?

  (A) 10 square units   (B) 20 square units

  (C) 24 square units   (D) 12 square units

23. A cube has a surface area of 96 cm^2. What is the edge length?

  (A) 3 cm   (B) 4 cm

  (C) 6 cm   (D) 8 cm

Find more at
ViewMath.com/NE-Grade6

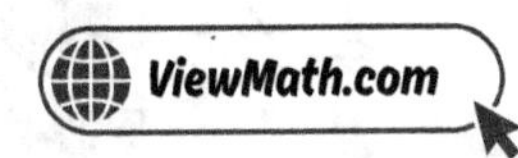

24. A circle has a diameter of 10 cm. What is its circumference? Use $\pi \approx 3.14$.

(A) 15.7 cm

(B) 31.4 cm

(C) 62.8 cm

(D) 314 cm

25. Look at the dot plot below.

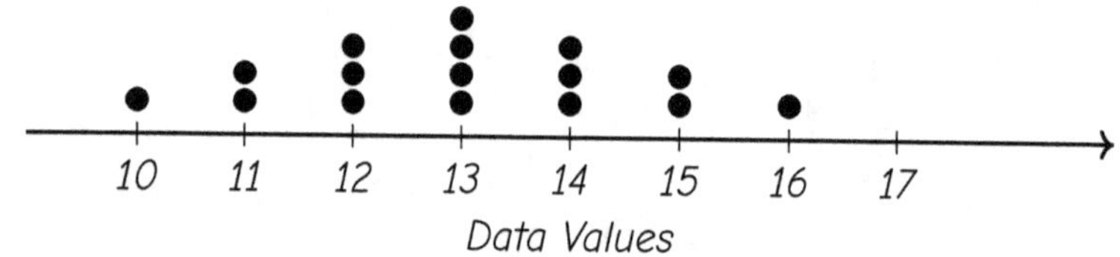

What is the shape of this data?

(A) Skewed to the right

(B) Skewed to the left

(C) Roughly symmetric

(D) Has a large gap in the middle

26. Four students' heights in inches are: $58, 60, 62, 64$. A fifth student who is 82 inches tall joins. What happens to the mean?

(A) It stays at 61.

(B) It increases to 65.2.

(C) It decreases.

(D) It increases to 82.

27. A number line below shows the positions of Q1, the median, and Q3 for a data set.

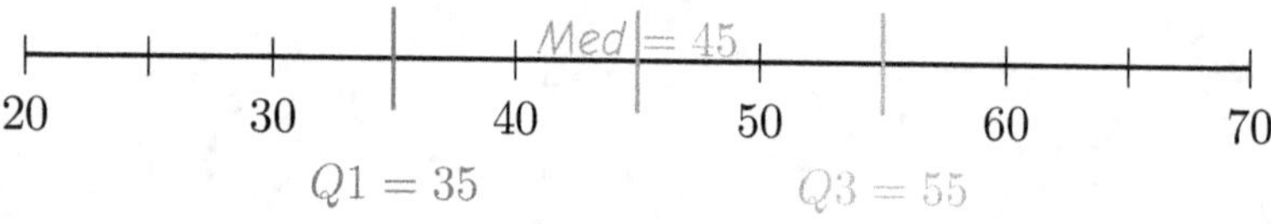

What is the IQR?

(A) 10

(B) 15

(C) 20

(D) 35

Find more at
ViewMath.com/NE-Grade6

28. Data (in order): $2, 4, 6, 8, 10, 12$. What is the median?

(A) 6

(B) 7

(C) 8

(D) 10

29. A report says: "The median score was 72 with an IQR of 15." What does this tell you?

(A) All scores are 72.

(B) The typical score is about 72, and the middle 50% of scores span 15 points.

(C) The highest score is 87.

(D) The lowest score is 57.

30. A box contains cards numbered 1 through 20. One card is drawn at random. What is the probability of drawing a number that is a multiple of 5? Write your answer as a fraction in simplest form.

Your Answer:

 # End of Practice Test 9 

Great job finishing the test!

 My Score

I got ___________ out of 30 questions right.

*Check your answers in the **Answer Key** at the back of the book.*

Review any questions you missed. That's how we learn!

Check Your Score Online!

Visit **ViewMath Academy** to enter your answers and see which topics you need to review. You can also explore lessons, take quizzes, track your scores, and save your progress!

viewmath.com/score/6.1.NE.24

Or go to viewmath.com/score and enter code: 6.1.NE.24

Practice Test 10

30 Questions

✏ Before You Start ✏

- ✔ **Read each question carefully** before choosing your answer.
- ✔ **Show your work** on scratch paper when you need to.
- ✔ **Skip hard questions** and come back to them later.
- ✔ **Check your answers** when you're done.
- ✔ **Take your time** — there's no rush!

⭐ You've Got This! ⭐

Do your best and show what you know!

1. The tape diagram below shows the ratio of apple juice to orange juice in a drink.

Apple: ▢▢▢

Orange: ▢▢▢▢▢

If each part equals 4 ounces, how many total ounces of drink are there?

(A) 12

(B) 20

(C) 32

(D) 8

2. The table below shows the prices at two stores for packs of pencils.

	Number of Pencils	Price
Store A	10	$4.50
Store B	8	$3.20

Which store has the lower unit price per pencil?

(A) Store A at $0.45 per pencil

(B) Store B at $0.40 per pencil

(C) Store A at $0.40 per pencil

(D) They have the same unit price.

3. Which pair of ratios are equivalent?

(A) 2 : 3 and 4 : 9

(B) 2 : 3 and 6 : 9

(C) 2 : 3 and 8 : 9

(D) 2 : 3 and 3 : 2

Get Online

Find more at
ViewMath.com/NE-Grade6

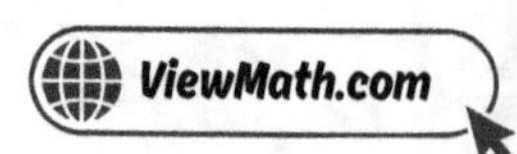

4. Which table matches a graph that passes through $(0, 0)$, $(2, 3)$, and $(4, 6)$?

(A)

x	y
2	3
6	9

(B)

x	y
2	3
6	8

(C)

x	y
3	2
6	9

(D)

x	y
2	4
6	12

5. A fence uses posts in a ratio of 1 post for every 6 feet. How many posts are needed for a 42-foot fence?

(A) 6

(B) 7

(C) 8

(D) 36

6. The number line below shows a distance in centimeters.

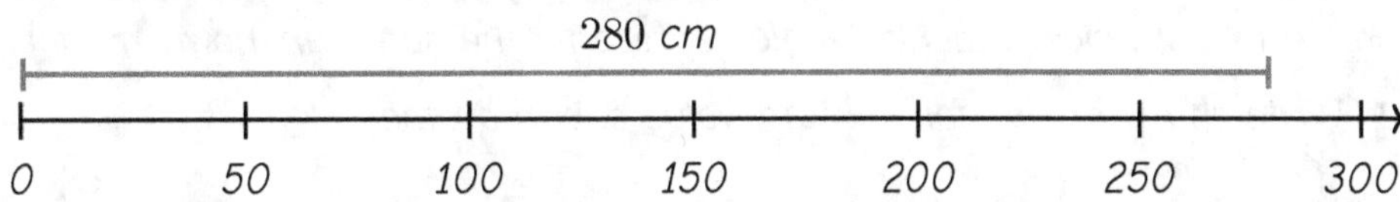

Part A: Convert 280 centimeters to meters.

Part B: Convert 280 centimeters to millimeters.

Part C: A doorway is 2 meters tall. Is 280 cm longer or shorter? By how much?

Your Answer:

7. Compute $3{,}780 \div 15$.

Your Answer:

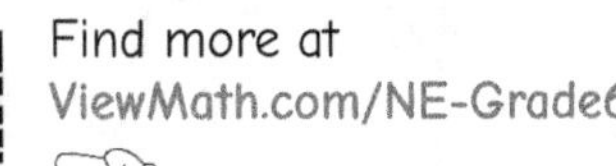

Find more at
ViewMath.com/NE-Grade6

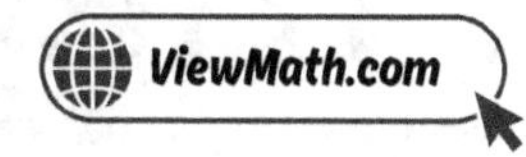

8. Compute $8.64 \div 0.4$.

Your Answer:

9. What is $|0|$?

(A) There is no answer

(B) -1

(C) 1

(D) 0

10. In which quadrant is the point $(2, -5)$ located?

(A) Quadrant I

(B) Quadrant II

(C) Quadrant III

(D) Quadrant IV

11. Look at the balance model below. Each triangle represents the same unknown number x, and each small square represents 1. Write an expression for the total value shown on the balance.

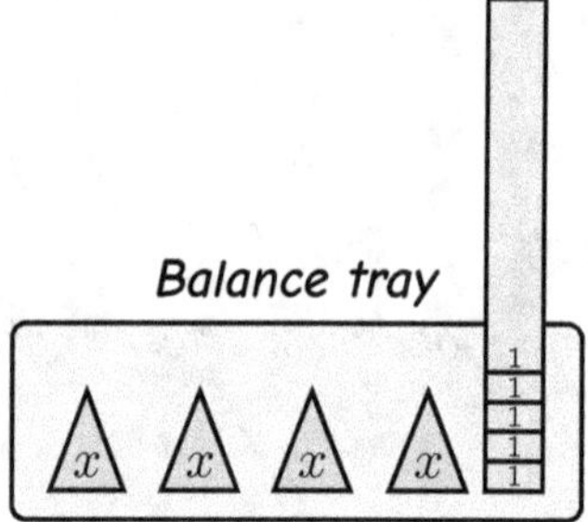

Your Answer:

Find more at
ViewMath.com/NE-Grade6

12. In the term $\dfrac{n}{5}$, what are the factors?

A) n and 5

B) n and $\dfrac{1}{5}$

C) 1 and n

D) 5 and $\dfrac{1}{n}$

13. The table below shows the cost of renting a bike using the expression $C = 8 + 3h$, where h is the number of hours. Fill in the missing values A and B.

Hours (h)	Cost (C)
1	$11
2	A
3	$17
5	B

Your Answer:

14. Which expression is equivalent to $7(k - 3) + 5$?

A) $7k - 16$

B) $7k + 2$

C) $7k - 26$

D) $7k - 21$

15. Solve: $7r = 63$

A) $r = 56$

B) $r = 70$

C) $r = 8$

D) $r = 9$

16. Write an inequality: You must be no fewer than 16 years old to drive. Let a = age.

Your Answer:

Find more at
ViewMath.com/NE-Grade6

ViewMath.com

17. Describe the graph of $n \leq 5$.

(A) Open circle at 5, shade left   (B) Closed circle at 5, shade left

(C) Open circle at 5, shade right   (D) Closed circle at 5, shade right

18. The table below shows how a savings account grows each week. Write the equation that relates savings s to weeks w, and predict the savings after 8 weeks.

Weeks (w)	Savings (s)
0	$20
1	$25
2	$30
3	$35
4	$40

Your Answer:

19. A triangle and a rectangle both have a base of 6 cm and a height of 10 cm. How do their areas compare?

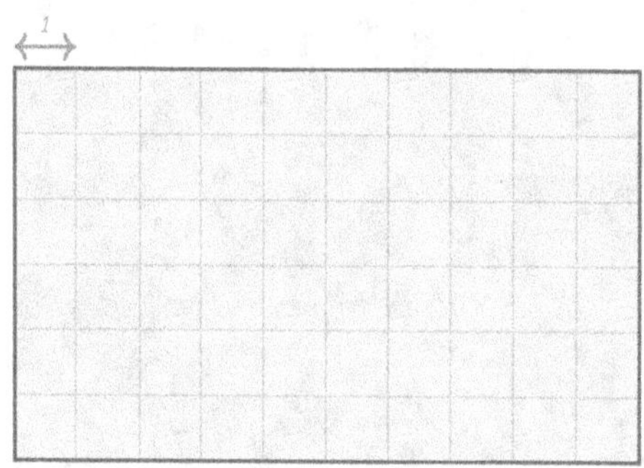

(A) The triangle has twice the area of the rectangle.   (B) They have the same area.

(C) The triangle has half the area of the rectangle.   (D) The rectangle has three times the area of the triangle.

20. A cube has edges of 4 cm. What is the volume?

(A) $12\ cm^3$ 　　　　　　　　　　　　　　(B) $16\ cm^3$

(C) $48\ cm^3$ 　　　　　　　　　　　　　　(D) $64\ cm^3$

21. What is the distance between the points $(2, 5)$ and $(2, -3)$?

(A) 2 units 　　　　　　　　　　　　　　(B) 5 units

(C) 8 units 　　　　　　　　　　　　　　(D) 3 units

22. An L-shaped figure has vertices $(0, 0)$, $(8, 0)$, $(8, 3)$, $(4, 3)$, $(4, 7)$, and $(0, 7)$. A student splits it into two rectangles. Which pair of rectangles correctly covers the figure?

(A) 8×7 and 4×3 　　　　　　　　(B) 8×3 and 4×4

(C) 4×7 and 4×3 　　　　　　　　(D) 8×3 and 4×7

23. A triangular prism has two equilateral triangle bases with side 6 cm and height 5.2 cm. The prism is 12 cm long. What is the total surface area?

Your Answer

24. Find the area of the semicircle shown below. Use $\pi \approx 3.14$.

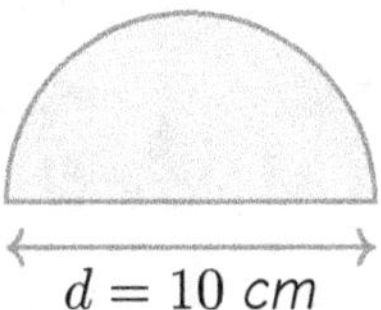

(A) $15.7\ cm^2$ 　　　　　　　　　　　　(B) $39.25\ cm^2$

(C) $78.5\ cm^2$ 　　　　　　　　　　　　(D) $157\ cm^2$

Find more at
ViewMath.com/NE-Grade6

25. *Student test times (minutes):* $8, 9, 10, 10, 11, 12, 25.$ *A classmate says the typical time is about* 12 *minutes. Is this a good estimate?*

(A) Yes, 12 is in the data set.

(B) Yes, because the outlier should be included.

(C) No, most times cluster around 9–11, so a typical value is closer to 10.

(D) No, the typical value is 25.

26. *A student scored 85, 90, 78, 92, and 95 on five tests. What score does the student need on a sixth test to have a mean of 90?*

Your Answer

27. *Data (in order):* $4, 6, 8, 10, 12, 14, 16.$ *What is Q1?*

(A) 4

(B) 6

(C) 8

(D) 10

28. Look at the box plot below.

What is the IQR of this data?

(A) 20

(B) 40

(C) 50

(D) 80

29. *When two data sets have very little overlap, what can you say?*

(A) The groups are very similar.

(B) The groups are clearly different from each other.

(C) Both groups have the same center.

(D) Both groups have the same spread.

Find more at
ViewMath.com/NE-Grade6

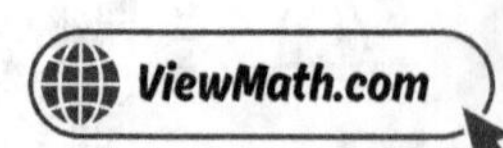

30. A bag holds 6 blue tokens, 4 red tokens, and 2 white tokens. What is the probability of drawing a blue token?

(A) $\dfrac{6}{10}$

(B) $\dfrac{1}{6}$

(C) $\dfrac{6}{12}$

(D) $\dfrac{4}{12}$

 # End of Practice Test 10 

Great job finishing the test!

 My Score

I got __________ out of 30 questions right.

*Check your answers in the **Answer Key** at the back of the book.*

Review any questions you missed. That's how we learn!

📊 Check Your Score Online!

Visit **ViewMath Academy** to enter your answers and see which topics you need to review. You can also explore lessons, take quizzes, track your scores, and save your progress!

viewmath.com/score/6.1.NE.25

Or go to viewmath.com/score and enter code: 6.1.NE.25

Answer Key & Explanations

Answer Key

First try each test on your own, then check your work here.

✓ Practice Test 1 — Answer Key

 5    C    C    C    C    8 full bottles    C   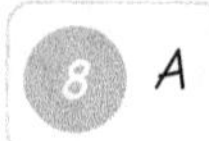 A    15

 $(-7, -3)$  B    $A = 4, B = (x + 3)$ 13. 22   14. B   15. $45   16. B

 No. Both shade to the left, but $x < 8$ has an open circle at 8 and $x \leq 8$ has a closed circle at 8.

 $y = 3$, $y = 11$, $y = 19$  C    B    D    B    B    C   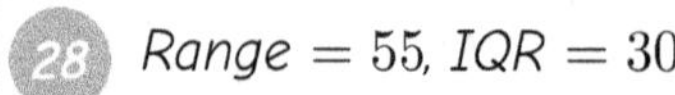 A

 A    B   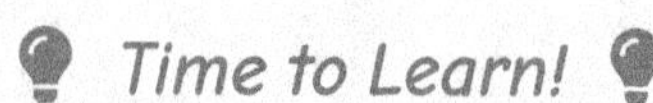 Range $= 55$, IQR $= 30$ 29. A   30. 0.625

💡 Time to Learn! 💡

Review the explanations below, **especially for the questions you missed**.

Understanding why each answer is correct builds stronger problem-solving skills.

Tip: Circle any questions you got wrong, then read their explanation carefully.

📖 Practice Test 1 — Detailed Explanations

1. *Each part $= 20 \div 4 = 5$. Vegetables $= 1 \times 5 = 5$.*

Find more at
ViewMath.com/NE-Grade6

2 From the graph, the truck travels 50 miles in 2 hours. Unit rate: $50 \div 2 = 25$ miles per hour.

3 From the double number line, 9 scoops aligns with 15 cups of water.

4 $20 \div 5 = 4$. The unit rate is 4 (units of y per 1 unit of x).

5 Scale: 2 cm = 15 km, so 1 cm = 7.5 km. For 5 cm: $5 \times 7.5 = 37.5$ km.

6 6 L = 6,000 mL. $6,000 \div 750 = 8$.

7 Total cans: $384 + 528 + 456 + 432 = 1,800$. Divide: $1,800 \div 12 = 150$ cans per classroom.

8 Multiply: 5.4×6. As whole numbers: $54 \times 6 = 324$. One decimal place: 32.4 liters.

9 $|-9| = 9$ and $|-6| = 6$. Adding these gives $9 + 6 = 15$. Find each absolute value first, then add.

10 Reflecting across the x-axis changes the sign of the y-coordinate: $(x, y) \rightarrow (x, -y)$. So $(-7, 3)$ becomes $(-7, -3)$.

11 Sandwiches cost $6s$ and drinks cost $2d$. Total: $6s + 2d$.

12 $4(x + 3) = 4 \times (x + 3)$. The two factors are 4 and $(x + 3)$.

13 $3(8 + 2) - 8 = 3(10) - 8 = 30 - 8 = 22$.

14 Distribute: $3 \times x + 3 \times 5 = 3x + 15$.

15 $\frac{g}{5} = 9$, so $g = 9 \times 5 = 45$ dollars.

Find more at
ViewMath.com/NE-Grade6

16 "Below 0" means less than 0: $t < 0$. (Not equal to 0, since it says "below.")

17 The direction is the same, but the circle type differs: open vs. closed.

18 $4(1) - 1 = 3.$ $4(3) - 1 = 11.$ $4(5) - 1 = 19.$

19 One sail: $\frac{1}{2} \times 3 \times 7 = 10.5 \ m^2$. Two sails: $10.5 \times 2 = 21 \ m^2$.

20 Box A volume $= 6 \times 4 \times 5 = 120.$ Box B: $120 = 10 \times 3 \times h = 30h$, so $h = 4.$

21 $|4 - (-5)| = |9| = 9$ units.

22 Base $= |2 - (-4)| = 6.$ Height $= |5 - 0| = 5.$ Area $= \frac{1}{2} \times 6 \times 5 = 15$ square units.

23 SA $= 2(10)(3) + 2(10)(7) + 2(3)(7) = 60 + 140 + 42 = 242 \ m^2$.

24 Radius $= 20 \div 2 = 10 \ ft.$ $A = \pi r^2 = 3.14 \times 10^2 = 3.14 \times 100 = 314 \ ft^2$. Choice A is the circumference. Choice D uses πd^2.

25 Right-skewed data has most values clustered on the lower end with a few high values stretching to the right.

26 Total students $= 15.$ Median is the 8th value. In order: three 1's (3 so far), five 2's (8 so far). The 8th value is 2.

27 Q1: median of $3, 5, 7 = 5.$ Q3: median of $10, 12, 14 = 12.$ IQR $= 12 - 5 = 7.$

28 Range $= 80 - 25 = 55.$ IQR $= 65 - 35 = 30.$

Find more at
ViewMath.com/NE-Grade6

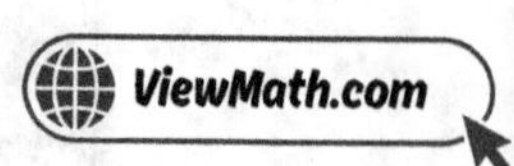

29 Class A's MAD is 3, meaning scores are typically within 3 of the mean. Class B's scores typically deviate 9 points. Class A's student is more likely to be near 75.

30 $P(blue) = \dfrac{3}{8}$. $P(not\ blue) = 1 - \dfrac{3}{8} = \dfrac{5}{8} = 0.625$.

✅ Practice Test 2 — Answer Key

1 C   **2** Part A: Lily earns \$12 per hour; Noah earns \$13 per hour.   Part B: Noah earns \$1 more per hour.

3 Part A: 2 : 1;   Part B: 4 cups of sugar   **4** C   **5** B   **6** C   **7** A   **8** \$16.43

9 B

10 The point $(9, 0)$ lies on the x-axis because its y-coordinate is 0. Points on the axes are not inside any quadrant.

11 C   **12** 5 and $(y + 8)$ **13** 30 square cm   **14** $6a + 10$ **15** B   **16** C   **17** B

18 C   **19** 54 in^2 **20** A   **21** A   **22** 64 square units   **23** C   **24** \$37.68   **25** D

26 Mean $\approx$ 66, Median $=$ 66   **27** C   **28** D   **29** B   **30** C

💡 Time to Learn! 💡

Review the explanations below, **especially for the questions you missed.**

Understanding why each answer is correct builds stronger problem-solving skills.

Tip: Circle any questions you got wrong, then read their explanation carefully.

📖 Practice Test 2 — Detailed Explanations

Find more at
ViewMath.com/NE-Grade6

1. Girls have 6 parts $= 24$, so each part $= 24 \div 6 = 4$. Boys have 4 parts $= 4 \times 4 = 16$.

2. Lily: $\$60 \div 5 = \$12/hr$. Noah: $\$52 \div 4 = \$13/hr$. Noah earns $\$13 - \$12 = \$1$ more per hour.

3. Part A: From the graph, $(2, 1)$ shows 2 cups of flour for every 1 cup of sugar. Part B: Following the pattern, $(8, 4)$, so 4 cups of sugar.

4. At $(2, 5)$: $x : y = 2 : 5$. Check: $(4, 10)$ simplifies to $2 : 5$ as well.

5. $360 \div 12 = 30$ minutes.

6. $1\ km = 1{,}000\ m = 100{,}000\ cm$. $10 \times 100{,}000 = 1{,}000{,}000\ cm$.

7. $7 \div 3 = 2$ R1; bring down $5 \rightarrow 15 \div 3 = 5$; bring down $6 \rightarrow 6 \div 3 = 2$. Answer: 252. Check: $252 \times 3 = 756$.

8. $12.95 + 3.48 = 16.43$. Hundredths: $5 + 8 = 13$, write 3, carry 1. Tenths: $9 + 4 + 1 = 14$, write 4, carry 1. Ones: $2 + 3 + 1 = 6$. Tens: 1.

9. Absolute value tells you how far a number is from zero. -12 is 12 units from zero, so $|-12| = 12$. A common mistake is to think $|-12| = -12$, but distance is never negative.

10. The four quadrants are the regions between the axes. A point is on an axis when $x = 0$ (on the y-axis) or $y = 0$ (on the x-axis). Since $(9, 0)$ has $y = 0$, it sits on the x-axis itself, which is the boundary between Quadrant I and Quadrant IV—not in either one.

11. Equally distributing c cupcakes into 6 boxes means dividing: $c \div 6$.

12. $5(y + 8) = 5 \times (y + 8)$. The factors are 5 and $(y + 8)$.

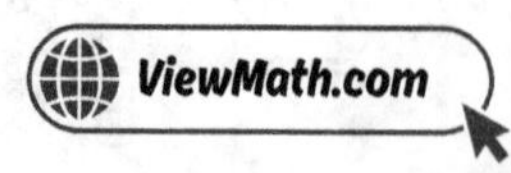

13 $A = \frac{1}{2} \times 10 \times 6 = \frac{1}{2} \times 60 = 30$ *square cm.*

14 $3a + 5a - 2a = 6a$ *and* $7 + 3 = 10$. *Result:* $6a + 10$.

15 *Check B:* $3(12) = 36$. ✓. *A gives* $x = 2$, *C gives* $x = 20$, *D gives* $x = 0$.

16 *"Fewer than 20" means less than 20:* $y < 20$.

17 *Both shade to the right. The only difference is the circle at 5: open for* $>$, *closed for* $\geq$.

18 *Each dog has 4 legs:* $L = 4d$.

19 *For a right triangle the legs are the base and height.* $A = \frac{1}{2} \times 9 \times 12 = 54$ in^2.

20 *Base area* $= 10 \times 4 = 40$. *Height* $= 120 \div 40 = 3$ *cm.*

21 *From* $(0,0)$ *with length 10 along the* x*-axis and width 4 along the* y*-axis, the opposite vertex is* $(10, 4)$.

22 *Side* $= |5 - (-3)| = 8$. *Area* $= 8 \times 8 = 64$ *square units.*

23 *A rectangular prism has 6 faces: 3 pairs of opposite rectangles.*

24 *Circumference* $= \pi d = 3.14 \times 6 = 18.84$ *ft. Cost* $= 18.84 \times 2 = \$37.68$.

25 *Most values are between 2 and 7, clustered around 4. The value 20 is far away from the rest, making it an outlier.*

26 *Values:* $60, 64, 64, 66, 68, 68, 72$. *Sum* $= 462$. *Mean* $= 462 \div 7 = 66$. *Median (4th of 7)* $= 66$.

Find more at
ViewMath.com/NE-Grade6

27) Range $= 25 - 8 = 17$.

28) Different data sets can have the same five-number summary. The five-number summary only captures five key values, not every data point.

29) Class X ranges from 25 to 40 (range = 15). Class Y ranges from 15 to 50 (range = 35). Class Y is much more spread out.

30) Event Y is positioned closest to 0.75 on the probability scale.

✅ Practice Test 3 — Answer Key

1 C	2 B	3 B	4 B	5 B	6 C	7 C	8 B	9 C	10 B
11 D	12 1	13 C	14 $8y + 24$	15 C	16 C	17 D	18 C	19 B	
20 B	21 B	22 28 square units	23 A	24 C	25 B				

26) Mean ≈ 13.3, Median $= 6$. The median is better.

27 B   28 B   29 C   30 C

💡 Time to Learn! 💡

Review the explanations below, **especially for the questions you missed.**

Understanding why each answer is correct builds stronger problem-solving skills.

Tip: Circle any questions you got wrong, then read their explanation carefully.

📖 Practice Test 3 — Detailed Explanations

Find more at
ViewMath.com/NE-Grade6

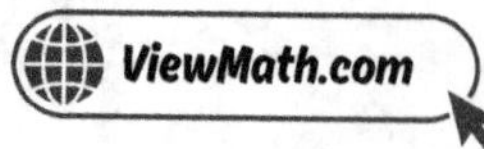

1. The question asks flour to eggs. Flour $= 3$, eggs $= 2$. The ratio is $3 : 2$.

2. $\$9 \div 12 = \$0.75$ per muffin.

3. $7 \times 3 = 21$ bananas, so $\$2 \times 3 = \$6$.

4. The ratio is $6 : 2 = 3 : 1$. When $x = 15$: $y = 15 \div 3 = 5$.

5. Unit rate: $\$24 \div 3 = \$8$ per shirt. 10 shirts: $10 \times \$8 = \$80$.

6. $3 \times 5{,}280 = 15{,}840$ feet.

7. $73 \div 24 = 3$ R1; bring down $4 \to 14 \div 24 = 0$ R14; bring down $4 \to 144 \div 24 = 6$. Answer: 306. The zero in the tens place must not be skipped. Check: $306 \times 24 = 7{,}344$.

8. Multiply as whole numbers: $12 \times 3 = 36$. Count decimal places: 1.2 has 1 and 0.3 has 1, total 2. Place the decimal: 0.36.

9. $|6| = 6$ and $|-6| = 6$. Both 6 and -6 are 6 units from zero, so both have an absolute value of 6.

10. In Quadrant II, the x-coordinate is negative (left of the y-axis) and the y-coordinate is positive (above the x-axis), giving the sign convention $(-, +)$.

11. Giving away 6 means subtracting: $x - 6$.

12. $4(a + b)$ is a single term — it is one product. (If you distribute it to $4a + 4b$, it would be 2 terms.)

13. $3^2 + 3 = 9 + 3 = 12$.

14 $8 \times y + 8 \times 3 = 8y + 24$.

15 Divide by 12: $p = 84 \div 12 = 7$.

16 "At most 21" means 21 or fewer: $d \leq 21$.

17 Closed circle at -2 (included) with shading to the left: $x \leq -2$.

18 $c = 7(6) = 42$ dollars.

19 $15 \times 4 = 60$ is the rectangle area. Sam forgot to multiply by $\frac{1}{2}$. The correct area is 30 cm^2.

20 $V = 3 \times 2 \times 2 = 12$ ft^3.

21 Vertical segment: $|5 - (-2)| = |7| = 7$ units.

22 Base $= 10 - 2 = 8$. Height $= 8 - 1 = 7$. Area $= \frac{1}{2} \times 8 \times 7 = 28$ square units.

23 $SA = 2(5)(3) + 2(5)(4) + 2(3)(4) = 30 + 40 + 24 = 94$ cm^2. Yes, it is correct.

24 The area of a circle is $A = \pi r^2$. Choices A and B are circumference formulas. Choice D incorrectly doubles the area formula.

25 Data set A range: $24 - 20 = 4$. Data set B range: $34 - 10 = 24$. Data set B is much more spread out.

26 Mean: $(3 + 4 + 5 + 6 + 7 + 8 + 60) \div 7 = 93 \div 7 \approx 13.3$. Median: 6. The outlier 60 inflates the mean. The median (6) is more typical.

Find more at
ViewMath.com/NE-Grade6

27 MAD (Mean Absolute Deviation) is calculated by finding the average of the distances of each data value from the mean.

28 $IQR = Q3 - Q1 = 40 - 20 = 20.$

29 Same medians, but Class A's IQR (6) is much smaller than Class B's (18). Class A's scores are more tightly clustered.

30 Total balls: $5 + 3 + 2 = 10.$ $P(yellow) = \dfrac{2}{10}.$ $P(not\ yellow) = 1 - \dfrac{2}{10} = \dfrac{8}{10}.$

✅ Practice Test 4 — Answer Key

1 B	2 3.5 hours	3 C	4 Line A. It has a rate of 6 per 1, which is higher than 4 per 1.

5 B	6 D	7 B	8 1.44	9 C	10 B	11 A	12 7 and 4	13 B

14 $11m + 2$	15 B	16 C	17 B	18 B	19 B	20 A	21 B	22 B

23 A	24 200.96 m^2	25 D	26 A	27 C	28 B

29 Period 3 performed better overall (mean 82 vs. 76, median 81 vs. 78) and was more consistent (IQR 8 vs. 16, range 18 vs.

30 C

💡 Time to Learn! 💡

Review the explanations below, **especially for the questions you missed.**

Understanding why each answer is correct builds stronger problem-solving skills.

Tip: Circle any questions you got wrong, then read their explanation carefully.

Find more at
ViewMath.com/NE-Grade6

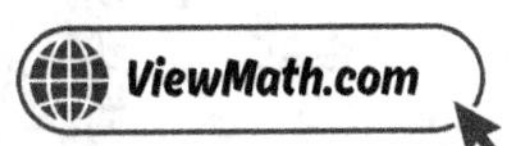

📖 Practice Test 4 — Detailed Explanations

1. "3 pencils for every 1 eraser" means pencils to erasers is $3 : 1$.

2. $49 \div 14 = 3.5$ hours.

3. The ratio is $4 : 10 = 2 : 5$. Row 3 should be $12 : 30$ (since $4 \times 3 = 12$ and $10 \times 3 = 30$), but it shows $12 : 25$.

4. Line A goes up 6 for every 1 unit right; Line B goes up 4. Higher rise per unit means a faster rate.

5. Unit rate: $180 \div 3 = 60$ mph. In 7 hours: $60 \times 7 = 420$ miles.

6. $3 \times 4 = 12$ cups.

7. The four steps of the standard algorithm are **Divide, Multiply, Subtract, Bring Down**. After dividing, you multiply the partial quotient by the divisor before subtracting.

8. Multiply as whole numbers: $36 \times 4 = 144$. Count decimal places: $1 + 1 = 2$. Place the decimal: 1.44. Check: $3.6 \times 0.4 = 1.44$

9. Distance from sea level is measured by absolute value. $|-40| = 40$ and $|25| = 25$. Since $40 > 25$, the diver is farther from sea level. The sign tells direction (above or below), not distance.

10. Quadrant II has sign convention $(-, +)$. Since $-4 < 0$ and $7 > 0$, the point $(-4, 7)$ is in Quadrant II. A common mistake is choosing Quadrant III, which has both coordinates negative.

11. $\frac{n}{3}$ is "n divided by 3." Adding 7 gives "7 more than n divided by 3."

12. 7 and 4 are the terms without variables, so they are constants.

Find more at
ViewMath.com/NE-Grade6

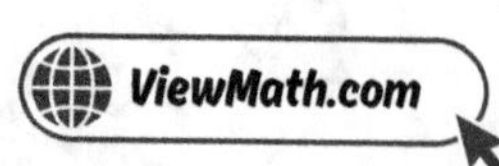

13 $P = 2(7) + 2(3) = 14 + 6 = 20$ cm.

14 Distribute: $8m + 2 + 3m$. Combine: $8m + 3m = 11m$. Result: $11m + 2$.

15 Subtract 15: $k = 32 - 15 = 17$.

16 "Below freezing" means below $32°F$: $t < 32$.

17 $x < -3$: $-4 < -3$ is true (shaded), but $-2 < -3$ is false (not shaded).

18 $21 = 3x$, so $x = 21 \div 3 = 7$.

19 $P = \frac{1}{2} \times 10 \times 6 = 30$. $Q = \frac{1}{2} \times 10 \times 8 = 40$. Difference $= 40 - 30 = 10$ cm^2.

20 $V = \frac{1}{2} \times \frac{1}{2} \times \frac{1}{2} = \frac{1}{8}$ m^3.

21 Horizontal side: $|7 - 1| = 6$. Vertical side: $|2 - (-4)| = 6$. Both sides are 6 units, so it is a square. The side length is 6.

22 Length $= |5 - (-1)| = 6$. Width $= |3 - (-2)| = 5$. Area $= 6 \times 5 = 30$ square units.

23 Net A is a valid cross-shaped cube net. Net B has two squares on the same side, which causes overlap. Net C forms a 2×3 block, which doesn't fold into a cube.

24 $A = \pi r^2 = 3.14 \times 8^2 = 3.14 \times 64 = 200.96$ m^2.

25 Most values are between 22 and 27. The value 45 is far from the rest, making it an outlier.

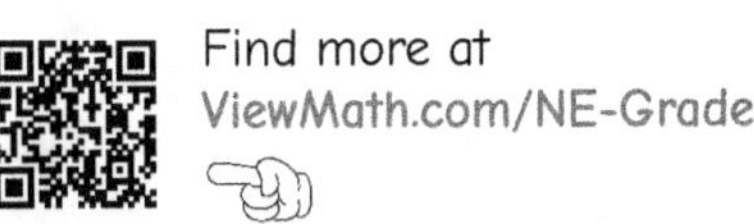
Find more at
ViewMath.com/NE-Grade6

26 Sum $= 119$. Mean $= 119 \div 7 = 17$. Median (4th of 7) $= 17$. They are equal because the data is symmetric.

27 Adding 5 to every value shifts the max and min by the same amount, so the difference (range) stays unchanged.

28 A long right whisker means some data values extend far above Q3, indicating the data is skewed to the right.

29 Period 3 has a higher center (both mean and median are higher) and a smaller spread (both range and IQR are smaller). This means Period 3 scored higher on average and their scores were more tightly clustered. Period 1's larger range (40) and IQR (16) suggest greater variability, possibly with some very low or very high scores.

30 $P(5) = \dfrac{1}{6} \approx 0.1\overline{6}$, which rounds to 0.17.

✅ Practice Test 5 — Answer Key

1 10   2 C   3 C   4 B   5 C   6 150 minutes; 9,000 seconds   7 C   8 C

9 B   10 $(3, -2)$ 11 $z - 9$ 12 D   13 25   14 B   15 C

16 $n = 4$ is NOT a solution to $n > 4$ (since 4 is not greater than 4). $n = 4$ IS a solution to $n \geq 4$ (since 4 equals 4).

17 B   18 B   19 C   20 B   21 20 units   22 B   23 A   24 B   25 B

26 D   27 2.4   28 B

29 Server A median $= \$45$, Server B median $= \$45$. Same typical earnings. Server B is more consistent (range $= \$1$

30 B

Find more at
ViewMath.com/NE-Grade6

💡 **Time to Learn!** 💡

Review the explanations below, **especially for the questions you missed**.

Understanding why each answer is correct builds stronger problem-solving skills.

Tip: Circle any questions you got wrong, then read their explanation carefully.

📖 Practice Test 5 — Detailed Explanations

1. Total parts $= 3 + 2 = 5$. Each part $= 25 \div 5 = 5$. Swimming $= 2 \times 5 = 10$.

2. $240 \div 8 = 30$ miles per gallon.

3. $4 \times 2.5 = 10$ loaves, so $6 \times 2.5 = 15$ cups. Alternatively, unit rate: $6 \div 4 = 1.5$ cups per loaf, then $1.5 \times 10 = 15$.

4. Ratio: $1 : 4$. When $x = 3$: $y = 3 \times 4 = 12$.

5. $8 \times 5 = 40$ km.

6. $2 \times 60 + 30 = 150$ minutes. $150 \times 60 = 9{,}000$ seconds.

7. $15 \div 5 = 3$; bring down $7 \to 7 \div 5 = 1$ R2; bring down $5 \to 25 \div 5 = 5$. Answer: 315. Check: $315 \times 5 = 1{,}575$.

8. 0.6 has 1 decimal place and 0.7 has 1 decimal place. Add them: $1 + 1 = 2$ decimal places. The product is 0.42.

9. $|-10| = 10$ because -10 is 10 units from zero. Absolute value is never negative. A common mistake is to think $|-10| = -10$, but distance from zero is always positive or zero.

Find more at
ViewMath.com/NE-Grade6

10   Start at $(0,0)$. Move 4 left and 3 up: $(-4, 3)$. Then move 7 right: $-4 + 7 = 3$, and 5 down: $3 - 5 = -2$. The final point is $(3, -2)$.

11   "9 fewer than z" means subtract 9 from z: $z - 9$.

12 $6(n + 2)$ means $6 \times (n + 2)$. The two factors being multiplied are 6 and $(n + 2)$.

13 $4^2 + 2(4) + 1 = 16 + 8 + 1 = 25$.

14   Distribute: $4 \times 3y - 4 \times 2 = 12y - 8$.

15   Divide both sides by 8: $w = 72 \div 8 = 9$.

16 $>$ does not include equality. $\geq$ includes equality.

17 $x > 2$ does NOT include 2, so the circle should be open, not closed.

18   Height increases by 3 cm per week: $h = 3w$.

19 $A = \frac{1}{2} \times 5 \times 12 = 30$ in^2.

20 $V = l \times w \times h$. If h becomes $2h$, then $V_{new} = l \times w \times 2h = 2(lwh)$, so the volume doubles.

21   Each side $= 5$. Perimeter $= 4 \times 5 = 20$ units.

22   Base $= |8 - 2| = 6$. Height $= |7 - 1| = 6$. Area $= \frac{1}{2} \times 6 \times 6 = 18$ square units.

23 $SA = 2(5)(5) + 2(5)(10) + 2(5)(10) = 50 + 100 + 100 = 250$ in^2.

Find more at
ViewMath.com/NE-Grade6

24 Pi (π) is defined as the ratio of a circle's circumference to its diameter. For every circle, $\pi = \dfrac{C}{d} \approx 3.14$.

25 The values 6 and 7 appear most frequently, and most of the data falls in the 6–7 range.

26 Mean $= \text{sum} \div 5 = 20$, so sum $= 20 \times 5 = 100$.

27 Distances from 10: $4, 2, 0, 2, 4$. $MAD = (4 + 2 + 0 + 2 + 4) \div 5 = 12 \div 5 = 2.4$

28 If the median is closer to $Q1$, the distance from median to $Q3$ is larger, meaning the upper half of the middle 50% is more spread out.

29 Server A data: $35, 40, 40, 45, 45, 45, 50, 50, 55, 60$. Median $= (45 + 45)/2 = 45$. Range $= 25$. Server B data: $40, 40, 40, 45, 45, 45, 45, 50, 50, 50$. Median $= (45 + 45)/2 = 45$. Range $= 10$. Same median but Server B is more consistent.

30 $\dfrac{2}{5} = 0.4$ and $\dfrac{3}{4} = 0.75$. Since $0.75 > 0.4$, Event B is more likely.

✅ Practice Test 6 — Answer Key

1 Bananas to yogurt: $3 : 4$;   Yogurt to bananas: $4 : 3$ **2** B   **3** B   **4** A   **5** B   **6** B

7 B   **8** C   **9** D   **10** C   **11** C   **12** C   **13** 16   **14** B   **15** C   **16** C

17 $x \le -4$ **18** $b = s \div 50$ (or $b = \dfrac{s}{50}$)   **19** B   **20** $60 \ cm^3$ **21** B   **22** 42 square units

23 A   **24** B   **25** A   **26** B   **27** D   **28** B   **29** B   **30** B

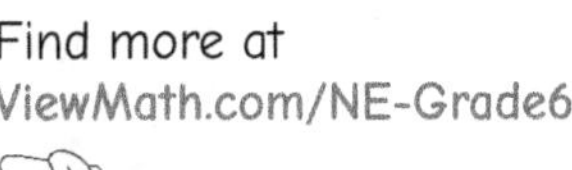
Find more at
ViewMath.com/NE-Grade6

> ### 💡 *Time to Learn!* 💡
>
> *Review the explanations below, **especially for the questions you missed**.*
>
> *Understanding why each answer is correct builds stronger problem-solving skills.*
>
> ***Tip:** Circle any questions you got wrong, then read their explanation carefully.*

📖 *Practice Test 6 — Detailed Explanations*

1 *"3 bananas for every 4 cups of yogurt" gives* $3 : 4$*. Flip the order for yogurt to bananas:* $4 : 3$*.*

2 *Brand X:* $\$8.75 \div 5 = \$1.75$ *each. Brand Y:* $\$4.50 \div 3 = \$1.50$ *each. Brand Y is cheaper.*

3 $2 \times 8 = 16$ *students, so* $3 \times 8 = 24$ *pencils.*

4 *Hours go on the x-axis, miles on the y-axis:* $(1, 30)$, $(2, 60)$, $(3, 90)$*.*

5 *"How much for one" is a unit rate problem. Divide to find the amount per 1 unit.*

6 *1 kg =* $1{,}000$ *g.* $7{,}000 \div 1{,}000 = 7$ *kg.*

7 $2{,}016 \div 14 = 144$ *Check:* $144 \times 14 = 2{,}016$*.*

8 *Move the decimal one place in both numbers:* $93.6 \div 12 = 7.8$*. Check:* $7.8 \times 1.2 = 9.36$*.*

9 *The opposite of* -3 *is 3. Both* -3 *and 3 are 3 units from zero, but on opposite sides of the number line.*

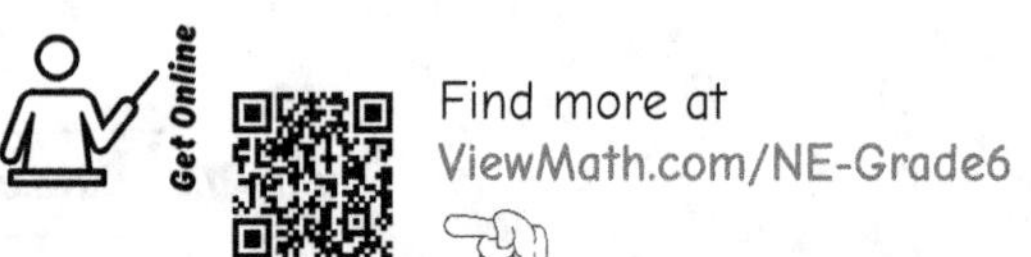

10 The x-coordinate -3 tells you to move 3 units to the left, and the y-coordinate 4 tells you to move 4 units up. Choice D swaps the coordinates.

11 Half of n is $\frac{n}{2}$. "6 more" means add 6: $\frac{n}{2} + 6$.

12 Part C points to 3, which is a term with no variable — a constant.

13 $8^2 = 64$. Then $64 \div 4 = 16$.

14 Distribute: $3x + 12 + 2x$. Combine: $3x + 2x = 5x$. Result: $5x + 12$.

15 Multiply by 6: $t = 5 \times 6 = 30$.

16 "More than \$25" means greater than 25: $d > 25$.

17 Closed circle means $\leq$ or $\geq$. Shading left means less than or equal to: $x \leq -4$.

18 Divide total students by 50 per bus: $b = s \div 50$.

19 $A = \frac{1}{2} \times 12 \times 9 = 54$ ft^2.

20 $V = 4 \times 3 \times 5 = 60$ cm^3. Each 1-cm cube has volume 1 cm^3.

21 $(5, -1)$ and $(5, 4)$ share the same x-coordinate, so the segment is vertical.

22 Length $= |9 - 2| = 7$. Width $= |5 - (-1)| = 6$. Area $= 7 \times 6 = 42$ square units.

23 $8 \times 5 \times 2 = 80$ is the volume. The actual surface area is $2(40) + 2(16) + 2(10) = 132$ cm^2.

Find more at
ViewMath.com/NE-Grade6

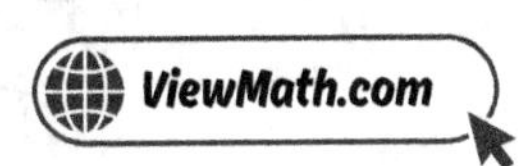

24 First find the radius: $r = 12 \div 2 = 6$ m. Then $A = \pi r^2 = 3.14 \times 6^2 = 3.14 \times 36 = 113.04$ m². Choice D uses the diameter instead of the radius in πd^2.

25 The outlier 100 increases the mean significantly. Without it, the center is around 17. With it, the mean jumps to about 27.

26 The outlier 50 increases the mean to about 15.1, which is higher than 6 of the 7 values. The median (10) better represents a typical value.

27 Range $= 100 - 1 = 99$. The outlier 100 makes the range very large.

28 Lower half: $3, 5, 7$. The median of the lower half is 5, so $Q1 = 5$.

29 The median and IQR are resistant to outliers and work better for skewed data. The mean and MAD are better for symmetric data.

30 Probability always ranges from 0 (impossible) to 1 (certain). It can never be negative or greater than 1.

✅ Practice Test 7 — Answer Key

1 6   2 B   3 16 and 20   4 C   5 $150 and $200   6 C   7 216   8 B

9 B   10 B   11 5 times a number x, plus 3 (or equivalent wording)   12 C   13 C   14 A

15 $m = 13$ 16 C   17 B   18 B   19 B   20 60 cm³   21 C   22 B

23 142 m²   24 C   25 Student A   26 C   27 Range $= 21$, IQR $= 5$. The IQR is better.

28 B   29 B   30 D

Find more at
ViewMath.com/NE-Grade6

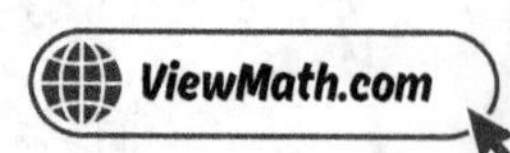

> 💡 **Time to Learn!** 💡
>
> Review the explanations below, **especially for the questions you missed.**
>
> Understanding why each answer is correct builds stronger problem-solving skills.
>
> **Tip:** Circle any questions you got wrong, then read their explanation carefully.

📖 Practice Test 7 — Detailed Explanations

1. $21 \div 7 = 3$, so multiply both by 3: iced tea $= 2 \times 3 = 6$.

2. $150 \div 5 = 30$ gallons per hour.

3. Row 2: $5 \times 2 = 10$, so $8 \times 2 = 16$. Row 3: $8 \times 4 = 32$, so $5 \times 4 = 20$.

4. $5 : 15$ simplifies to $1 : 3$.

5. Total parts $= 7$. Each part $= \$350 \div 7 = \$50$. Family 1: $3 \times \$50 = \$150$. Family 2: $4 \times \$50 = \$200$.

6. 1 foot $= 12$ inches. $5 \times 12 = 60$ inches.

7. $90 \div 42 = 2$ R6; bring down 7 $\to$ $67 \div 42 = 1$ R25; bring down 2 $\to$ $252 \div 42 = 6$. Answer: 216. Check: $216 \times 42 = 9{,}072$.

8. Multiply: 2.75×3. As whole numbers: $275 \times 3 = 825$. Two decimal places: $\$8.25$.

9. The opposite of a number is the same distance from zero on the other side of the number line. The opposite of 5 is -5 because both are 5 units from zero.

Find more at
ViewMath.com/NE-Grade6

10 When the y-coordinate is 0, the point lies on the x-axis. The point $(-6, 0)$ is 6 units to the left of the origin on the x-axis. A common mistake is thinking a negative x-value means the point is on the y-axis.

11 $5x$ means "5 times x" and $+3$ means "plus 3."

12 $5m - 2n + 7 + m$ has 4 terms: $5m$, $2n$, 7, and m.

13 $2(9) + 2(4) = 18 + 8 = 26$.

14 Like terms: $5p + 2p + 4p = 11p$. Constants: $3 - 1 = 2$. Result: $11p + 2$.

15 Subtract 17: $m = 30 - 17 = 13$.

16 "At most 50" means 50 or less: $w \leq 50$.

17 $>$ does not include 4, so use an open circle to show 4 is NOT a solution.

18 $y = 4(3) = 12$. The point $(3, 12)$ satisfies $y = 4x$.

19 The height is the perpendicular distance from the base to the opposite vertex. It is not always a side of the triangle.

20 $V = 2.5 \times 4 \times 6 = 60$ cm^3.

21 The base goes from $(0, 0)$ to $(8, 0)$. Distance $= |8 - 0| = 8$ units.

22 Side $= |4 - (-2)| = 6$. Area $= 6 \times 6 = 36$ square units.

Find more at
ViewMath.com/NE-Grade6

23. $SA = 2(7)(5) + 2(7)(3) + 2(5)(3) = 70 + 42 + 30 = 142\ m^2.$

24. Area of $P = \pi \times 4^2 = 16\pi.$ Area of $Q = \pi \times 8^2 = 64\pi.$ Ratio $= 64\pi \div 16\pi = 4.$ When the radius doubles, the area quadruples because area depends on $r^2.$

25. A smaller range means the data values are closer together. Student A's data is more consistent because the values span only 5 units.

26. Total $= 15 \times 6 = 90.$ Remove 9: $90 - 9 = 81.$ New mean $= 81 \div 5 = 16.2.$

27. Range $= 30 - 9 = 21.$ Q1: median of $9, 10, 10 = 10.$ Q3: median of $14, 15, 30 = 15.$ IQR $= 15 - 10 = 5.$ The $30 wage is an outlier that makes the range misleadingly large. The IQR (5) better represents how spread out most wages are.

28. The five-number summary consists of minimum, first quartile (Q1), median, third quartile (Q3), and maximum.

29. The range is 12 and the data is evenly spaced. Whether "very spread out" is accurate depends on context, but the data is quite orderly and not wildly spread.

30. $P(\text{yellow}) = \dfrac{2}{8} = \dfrac{1}{4} = 0.25.$

✓ Practice Test 8 — Answer Key

1. B

2. Part A: 0.5 laps per minute (or 1 lap every 2 minutes). Part B: 7 laps.

3. 15

4. B

5. 40 minutes

6. C

7. A

8. B

9. C

10. Part A: Quadrant II; Part B: $(8, 1)$, Quadrant I; Part C: $(-8, -1)$, Quadrant III

11. B

12. B

Find more at
ViewMath.com/NE-Grade6

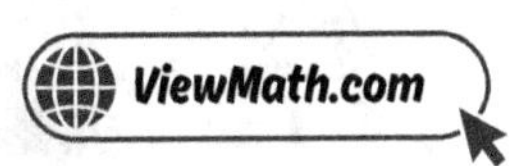

13 $23   **14** $6x + 12$ **15** $n - 19 = 31; n = 50$ **16** $g \le 4$

17 $x = 3$ IS a solution to $x \le 3$ because $3 \le 3$ is true. $x = 3$ is NOT a solution to $x < 3$ because $3 < 3$ is false.

18 C   **19** $13.5\ m^2$ **20** B   **21** $(9, -5)$ **22** B   **23** B   **24** $254.34\ in^2$

25 Skewed to the left (or: most data on the right with a tail to the left).   **26** C   **27** 6   **28** C

29 Class A performed better (higher median) and was more consistent (smaller IQR).   **30** B

💡 Time to Learn! 💡

Review the explanations below, **especially for the questions you missed**.

Understanding why each answer is correct builds stronger problem-solving skills.

Tip: Circle any questions you got wrong, then read their explanation carefully.

📖 Practice Test 8 — Detailed Explanations

1 Dogs have 4 parts. Each part $= 2$ animals. Dogs $= 4 \times 2 = 8$.

2 Part A: From the graph, the swimmer completes 1 lap every 2 minutes, so the rate is $\frac{1}{2}$ lap per minute. Part B: $14 \div 2 = 7$ laps.

3 $2 \times 5 = 10$ strawberries, so $3 \times 5 = 15$ cups of yogurt.

4 x doubled from 4 to 8, so y doubles from 7 to 14.

5 $1,000 \div 250 = 4$ batches. $10 \times 4 = 40$ minutes.

Find more at
ViewMath.com/NE-Grade6

6 1 yard $= 3$ feet. $7 \times 3 = 21$ feet.

7 To check a division, multiply the quotient by the divisor. If $4{,}752 \div 12 = 396$, then 396×12 should equal $4{,}752$.

8 Multiply as whole numbers: $12 \times 5 = 60$. Count decimal places: 0.12 has 2 and 0.5 has 1, total 3. Place the decimal: $0.060 = 0.06$.

9 The opposite of 0 is 0 itself. Zero is neither positive nor negative and sits right in the middle of the number line. It is the only number that is its own opposite.

10 Part A: $(-8, 1)$ has signs $(-, +)$, which is Quadrant II. Part B: Reflecting across the y-axis flips the x-sign: $(-8, 1) \to (8, 1)$. Signs $(+, +)$ is Quadrant I. Part C: Reflecting across the x-axis flips the y-sign: $(-8, 1) \to (-8, -1)$. Signs $(-, -)$ is Quadrant III.

11 "8 less than a number" means subtract 8 from n: $n - 8$.

12 The term d means $1 \cdot d$, so the coefficient is 1. (The subtraction sign belongs to the operation, not the coefficient of d itself in the original expression.)

13 $5 + 3(6) = 5 + 18 = 23$ dollars.

14 Perimeter $= 2(2x + 5) + 2(x + 1) = 4x + 10 + 2x + 2 = 6x + 12$.

15 Add 19: $n = 31 + 19 = 50$.

16 "No more than 4" means 4 or fewer. $g \leq 4$.

17 $\leq$ includes the boundary value; $<$ does not.

18 You choose x (independent), and y is computed from it (dependent). y increases as x increases.

19 $A = \frac{1}{2} \times 4.5 \times 6 = 13.5 \ m^2$.

20 Box A: $V = 5 \times 2 \times 3 = 30$. Box B: $V = 3 \times 3 \times 4 = 36$. Box B is greater.

21 The fourth vertex must share an x-value with $(9, 3)$ and a y-value with $(2, -5)$, giving $(9, -5)$.

22 Base $= 8$, height $= 6$. Area $= \frac{1}{2} \times 8 \times 6 = 24$ square units.

23 A net is a flat pattern that, when folded, forms a 3D shape. Each section of the net becomes a face.

24 First find the radius: $r = 18 \div 2 = 9$ in. Then $A = \pi r^2 = 3.14 \times 9^2 = 3.14 \times 81 = 254.34 \ in^2$.

25 The peak is at 5, with values trailing to the left toward 1. Most data is on the higher end.

26 Most values are between 2 and 5, but the outlier 100 increases the sum significantly, pulling the mean up to 22.8.

27 Distances from 60: $10, 5, 0, 5, 10$. $MAD = (10 + 5 + 0 + 5 + 10) \div 5 = 30 \div 5 = 6$.

28 A larger IQR means the middle 50% of the data covers a wider range. Class B's middle scores are more spread out.

29 Class A's median is 8 points higher. Class A's IQR is 12 points smaller, meaning its middle 50% is more tightly clustered.

30 Numbers ≤ 3: $1, 2, 3$. That is 3 out of 6. $P(\leq 3) = \frac{3}{6} = \frac{1}{2}$.

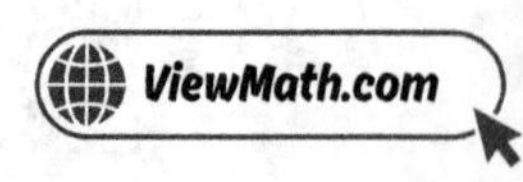

✅ Practice Test 9 — Answer Key

 C  B  B  B  C  C  B

 Part A: 27.54 km;   Part B: 9.18 km per day    D

 Part A: $A = (2, 5)$, $B = (5, 1)$, $C = (2, 1)$;   Part B: $A' = (-2, 5)$, $B' = (-5, 1)$, $C' = (-2, 1)$;   Part C: Quadrant II

 B  B  A  A  D  C  D  C  C  B

 A  C  B  B  C  B  C  B 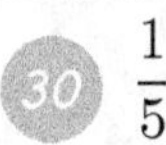 B  $\dfrac{1}{5}$

💡 Time to Learn! 💡

Review the explanations below, **especially for the questions you missed**.

Understanding why each answer is correct builds stronger problem-solving skills.

Tip: Circle any questions you got wrong, then read their explanation carefully.

📖 Practice Test 9 — Detailed Explanations

1. Total parts $= 1 + 3 + 2 = 6$. Each part $= 18 \div 6 = 3$. Blue $= 3 \times 3 = 9$.

2. Divide: $120 \div 4 = 30$ pages per minute.

3. $3 \times 2 = 6$, so $5 \times 2 = 10$.

4. $1 : 3$ multiplied by 3 gives $3 : 9$, so $(3, 9)$ is on the line.

Find more at
ViewMath.com/NE-Grade6

5 $3 \times 4 = 12$ *raisins, so* $4 \times 4 = 16$ *cups of granola.*

6 $1\ L = 1,000\ mL$. $2.5 \times 1,000 = 2,500\ mL$.

7 $26 \div 16 = 1$ *R10; bring down* $8 \to 108 \div 16 = 6$ *R12; bring down* $8 \to 128 \div 16 = 8$. *Answer:* 168. *Check:* $168 \times 16 = 2,688$.

8 *Part A:* $8.76 + 12.30 + 6.48 = 27.54\ km$. *Part B:* $27.54 \div 3 = 9.18\ km$ *per day. Check:* $9.18 \times 3 = 27.54$.

9 *A number and its opposite are the same distance from zero on different sides, so they cancel each other out. For example,* $6 + (-6) = 0$ *and* $-9 + 9 = 0$.

10 *Part A: Reading from the grid,* $A = (2, 5)$, $B = (5, 1)$, $C = (2, 1)$. *Part B: Reflecting across the* y*-axis changes the sign of each* x*-coordinate:* $A' = (-2, 5)$, $B' = (-5, 1)$, $C' = (-2, 1)$. *Part C: All reflected vertices have negative* x*-coordinates and positive* y*-coordinates, which is the sign convention* $(-, +)$ *for Quadrant II.*

11 *The product of 4 and t is* $4t$. *The sum of 9 and that product is* $9 + 4t$.

12 *The terms are* $4x$, 9, *and* $2y$. *Terms are separated by* $+$ *or* $-$ *signs.*

13 $6(4) - 4^2 = 24 - 16 = 8$.

14 $7n$ *and* $4n$ *are like terms. Add coefficients:* $7 + 4 = 11$, *so* $7n + 4n = 11n$.

15 *Division is undone by multiplication. Multiply both sides by 3:* $n = 12 \times 3 = 36$.

16 $n \leq 8$ *means* n *is 8 or less.* $8 \leq 8$ *is true. 8.5, 9, and 10 are all greater than 8.*

17 $10 < 10$ *is false. 10 is NOT less than 10, so it is not a solution.*

Find more at
ViewMath.com/NE-Grade6

18 $d = 60(3) = 180$ *miles.*

19 $A = \frac{1}{2} \times 20 \times 7 = 70$ *in*^{2}.

20   *Base area* $= 6 \times 4 = 24$ *in*^{2}. *Height* $= 72 \div 24 = 3$ *in.*

21   *From* $(-2, -2)$, *moving 5 right gives* $x = 3$, *and 5 up gives* $y = 3$. *The opposite vertex is* $(3, 3)$.

22   *Length* $= 6$, *width* $= 4$. *Area* $= 6 \times 4 = 24$ *square units.*

23 $6s^2 = 96$, *so* $s^2 = 16$, *giving* $s = 4$ *cm.*

24 $C = \pi d = 3.14 \times 10 = 31.4$ *cm. Choice A uses* πr *incorrectly as the full circumference, and choice C doubles the correct answer.*

25   *The dot plot shows a peak at 13 with roughly equal numbers of dots on each side. This is approximately symmetric.*

26   *Original mean:* $(58 + 60 + 62 + 64) \div 4 = 61$. *New mean:* $(58 + 60 + 62 + 64 + 82) \div 5 = 326 \div 5 = 65.2$.

27 $IQR = Q3 - Q1 = 55 - 35 = 20$.

28   *Even number of values. The two middle values are 6 and 8. Median* $= (6 + 8) \div 2 = 7$.

29   *The median (72) is the typical value. The IQR (15) tells you the middle half of the data spans 15 points.*

30   *Multiples of 5 from 1 to 20:* $5, 10, 15, 20$. *That is 4 favorable outcomes out of 20.* $P = \dfrac{4}{20} = \dfrac{1}{5}$.

✔ Practice Test 10 — Answer Key

1 C   **2** B   **3** B   **4** A   **5** C

6 Part A: 2.8 m;   Part B: 2,800 mm;   Part C: Longer by 0.8 m (80 cm).   **7** 252   **8** 21.6   **9** D

10 D   **11** $4x + 5$ **12** B   **13** A = \$14, B = \$23   **14** A   **15** D   **16** $a \geq 16$ **17** B

18 $s = 5w + 20$; after 8 weeks: \$60   **19** C   **20** D   **21** C   **22** B   **23** 247.2 cm^2 **24** B

25 C   **26** 100   **27** B   **28** B   **29** B   **30** C

💡 Time to Learn! 💡

Review the explanations below, **especially for the questions you missed**.

Understanding why each answer is correct builds stronger problem-solving skills.

Tip: Circle any questions you got wrong, then read their explanation carefully.

📖 Practice Test 10 — Detailed Explanations

1 Apple has 3 parts, orange has 5 parts. Total parts = 8. Total ounces = $8 \times 4 = 32$.

2 Store A: $\$4.50 \div 10 = \$0.45$. Store B: $\$3.20 \div 8 = \$0.40$. Store B is cheaper per pencil.

3 $2 : 3$ multiplied by 3 gives $6 : 9$. The other pairs do not simplify to $2 : 3$.

4 The ratio is $2 : 3$. Choice A continues with $6 : 9 = 2 : 3$. Choice B has $6 : 8$ which is not $2 : 3$.

Find more at
ViewMath.com/NE-Grade6

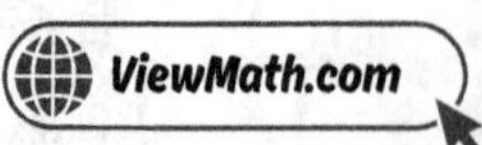

5 $42 \div 6 = 7$ sections, which needs $7 + 1 = 8$ posts (one at each end). Note: if counting "per 6 feet" as a rate, $42 \div 6 = 7$, but fences need a post at the start too, giving 8.

6   Part A: $280 \div 100 = 2.8$ m. Part B: $280 \times 10 = 2{,}800$ mm. Part C: $2.8 - 2 = 0.8$ m longer.

7 $37 \div 15 = 2$ R7; bring down 8 $\rightarrow$ $78 \div 15 = 5$ R3; bring down 0 $\rightarrow$ $30 \div 15 = 2$. Answer: 252. Check: $252 \times 15 = 3{,}780$.

8   Move the decimal one place in both numbers: $86.4 \div 4 = 21.6$. Check: $21.6 \times 0.4 = 8.64$.

9   The absolute value of 0 is 0 because zero is 0 units from itself. Zero is the only number whose absolute value is 0.

10   Quadrant IV has sign convention $(+, -)$. Since $2 > 0$ and $-5 < 0$, the point $(2, -5)$ is in Quadrant IV. A common error is confusing Quadrant IV with Quadrant II which has signs $(-, +)$.

11   There are 4 triangles (each worth x) and 5 unit squares. The expression is $4x + 5$.

12 $\frac{n}{5} = \frac{1}{5} \times n$. The factors are $\frac{1}{5}$ and n.

13 $h = 2$: $8 + 3(2) = 14$. $h = 5$: $8 + 3(5) = 23$.

14   Distribute: $7k - 21 + 5$. Combine: $-21 + 5 = -16$. Result: $7k - 16$.

15   Divide by 7: $r = 63 \div 7 = 9$.

16   "No fewer than 16" means 16 or more: $a \geq 16$.

17 $\leq$ means 5 is included (closed circle). Less than 5 is to the left (shade left).

Find more at
ViewMath.com/NE-Grade6

18 Savings increase by \$5 per week, starting at \$20: $s = 5w + 20$. When $w = 8$: $5(8) + 20 = 60$.

19 Rectangle area $= 6 \times 10 = 60$ cm^2. Triangle area $= \frac{1}{2} \times 6 \times 10 = 30$ cm^2. The triangle is half.

20 A cube has $l = w = h = 4$. $V = 4 \times 4 \times 4 = 64$ cm^3.

21 Same x-coordinate, so it is a vertical distance. $|5 - (-3)| = |8| = 8$ units.

22 Bottom rectangle: $8 \times 3 = 24$. Left rectangle above: $4 \times (7 - 3) = 4 \times 4 = 16$. Total: 40. This matches option B.

23 2 triangles: $2 \times \frac{1}{2}(6)(5.2) = 31.2$. 3 rectangles: $3 \times 6 \times 12 = 216$. Total: $31.2 + 216 = 247.2$ cm^2.

24 Radius $= 10 \div 2 = 5$ cm. Full circle area $= \pi r^2 = 3.14 \times 25 = 78.5$ cm^2. Semicircle area $= 78.5 \div 2 = 39.25$ cm^2. Choice C is the full circle area.

25 Most data values are between 8 and 12, clustering around 10. The 25 is an outlier. A typical time is about 10 minutes.

26 Need total $= 90 \times 6 = 540$. Current total $= 85 + 90 + 78 + 92 + 95 = 440$. Need $540 - 440 = 100$.

27 Lower half: $4, 6, 8$. The median of the lower half is 6, so $Q1 = 6$.

28 $Q1 = 30$, $Q3 = 70$. $IQR = 70 - 30 = 40$.

29 Little overlap means most values in one group are different from the values in the other group, which signals a clear difference between the groups.

30 Total tokens: $6 + 4 + 2 = 12$. $P(\text{blue}) = \dfrac{6}{12} = \dfrac{1}{2}$.

Well done checking your answers!

Keep practicing to strengthen your skills.

Author's Final Note

I hope you enjoyed this book as much as I enjoyed writing it. Whether you are a student working through the material, a parent supporting your child's learning, or a teacher guiding your class, I have tried to make this book as clear and engaging as possible. I hope I have succeeded. If you have any suggestions for improvement, please let me know. I would love to hear from you.

The accuracy of calculations is very important to me. We have done our best, but I also expect that I have made some minor errors. Constant improvement is the name of the game. If you find any errors, please let me know. I will fix them in the next edition.

For students: Your learning journey does not end here. I have written a series of books to help you learn math. Make sure you browse through them. I especially recommend workbooks and practice tests to help you prepare for your exams.

For parents: Thank you for investing in your child's education. I encourage you to explore the companion resources available online to help support your child outside the classroom.

For teachers: Thank you for the invaluable work you do every day. I hope this book serves as a useful resource in your classroom. Feel free to reach out if you have suggestions or would like to discuss how best to use this book with your students.

I also enjoy reading your reviews. If you have a moment, please leave a review on where you found this book. It will help others find this book. If you have any questions or comments, please feel free to contact me at drNazari@ViewMath.com.

And one last thing: Remember to use online resources for additional help. I recommend using the resources on https://ViewMath.com You can find video lessons, practice problems, and more. You can also use the online companion for this book to track your progress and access additional resources.

Wishing all students the best in their studies, parents every success in supporting their children, and teachers continued inspiration in their classrooms!

Dr. A. Nazari

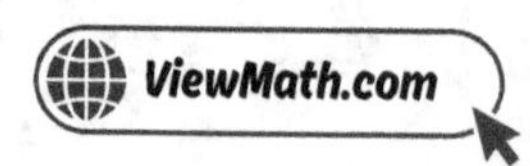

Great Job! Keep Learning with ViewMath!

Keep up the great work! Visit **viewmath.com/NE-Grade6** for free lessons, quizzes, and more.

Study Guide

Workbook

Step-by-Step

Grade 6 3 Practice Tests
Nebraska
3 Practice Tests

5 Practice Tests

7 Practice Tests

Find more at
ViewMath.com/NE-Grade6

ViewMath.com